Contents: delivering the EDEXCEL GCSE Graphic Products Specification

Welcome to Edexcel GCSE Graphic Products

Why should I choose GCSE Graphic Products?

Because you will:

- think creatively
- solve problems
- design your own product of the future
- make models
- test your ideas.

What will I learn?

The course covers an exciting range of products including packaging, point-of-sale display, interior and garden design and 3D product (concept) design.

Over the course of the two years you will develop a whole range of creative designing and making skills, technical knowledge and understanding relating to graphic products. There are two units: Your controlled assessment and Knowledge and Understanding of Graphic Products.

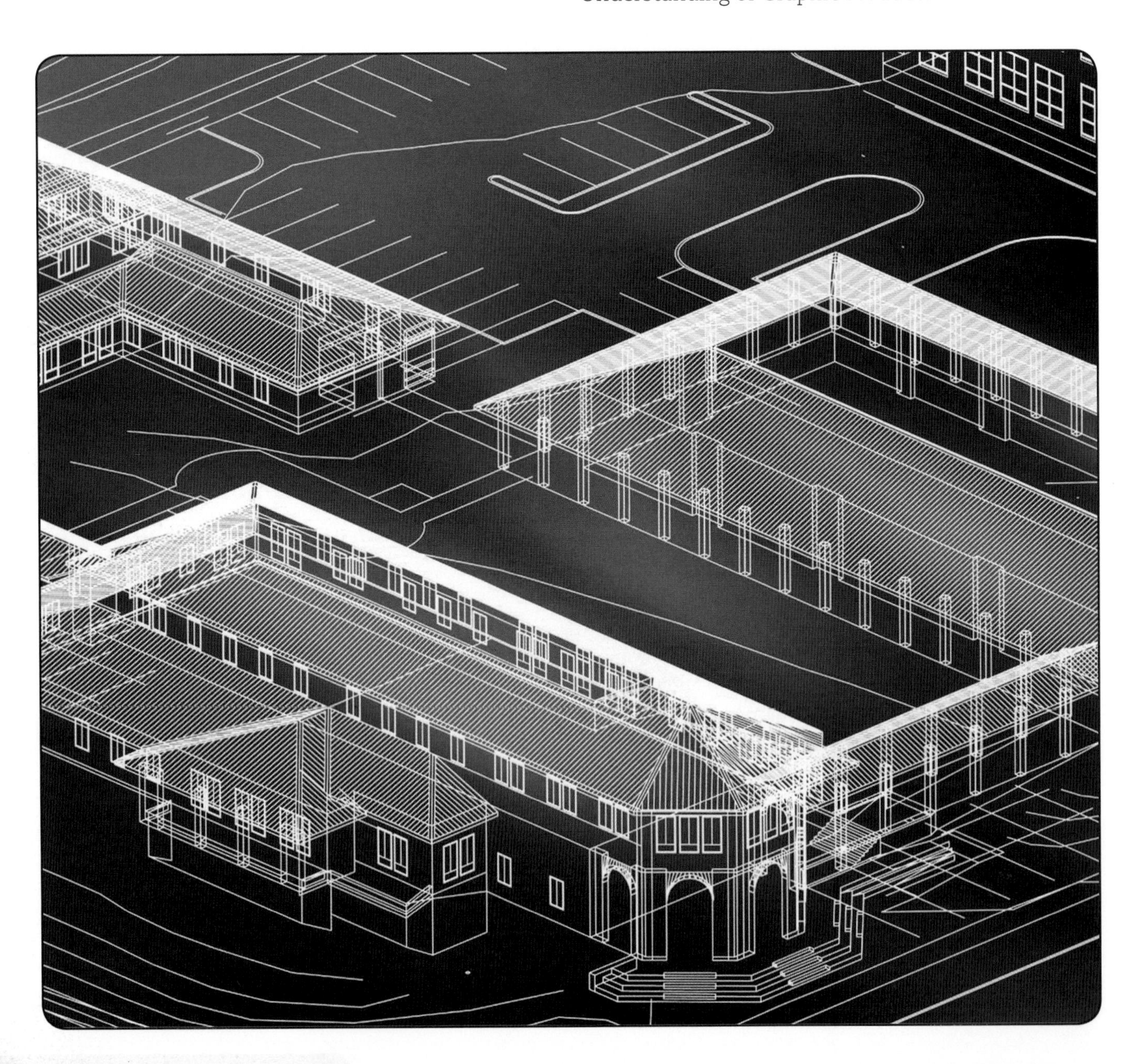

Unit 1: Your controlled assessment

This is the unit where you can really get stuck in! You can either create a combined design and make activity, where you design a product and then make a model of it, or you can complete separate design and make activities, where you design one product and make another.

Unit 2: Knowledge and Understanding of Graphic Products

This unit focuses on developing your knowledge and understanding of a wide range of materials and processes used in the field of design and technology. You will learn about industrial and commercial practices and the importance of quality checks, as well as the health and safety issues that have to be considered. What you learn in this unit will be applied during your controlled assessment Unit 1 creative design and make activities.

How will I be assessed?

Unit 1 Creative Design and Make Activities is the controlled assessment (coursework) unit. This means that your work will be internally assessed by your tutor. It is worth **60 per cent** of your overall course.

For Unit 2: Knowledge and Understanding of Graphic Products, you will sit a 1 hour 30 minutes examination that is assessed by Edexcel. It is worth **40 per cent** of your overall course.

The great thing about the course is that each of the units can be re-taken once. This means that, if you don't achieve the mark you wanted, then you can have another go!

About this book

This book is design to make learning about graphic products accessible and interesting. Look out for these features as you use your book.

Objectives provide a clear overview of what you will learn in the section.

Clear and accessible diagrams highlight key concepts.

Apply it! features help relate key content to the controlled assessment activities.

Forming techniques

Objectives

- **Describe** the processes of blow moulding, injection moulding, vacuum forming and line bending thermoplastics.
- **Describe** quality control (QC) inspection and testing procedures when thermoforming products.
- **Describe** the process of printing directly onto thermoformed products.
- **Explain** the advantages and disadvantages of using these methods for batch and mass production of graphic products.

Thermoforming products

Many products are batch- and mass-produced using thermoforming techniques to mould and shape polymers. Thermoplastics are commonly used because they can be easily moulded and any waste produced can be recycled and used again in the process.

Blow moulding and injection moulding can produce large quantities of identical products very quickly. Vacuum forming is also widely used in industry and is ideal for producing batches of similar products within schools. In the school workshop, line bending of acrylic enables you to produce high-quality products or models with some accuracy.

Blow moulding

In the blow-moulding process, a hollow thermoplastic tube or parison is extruded or forced out between a split two-piece mould and clamped at both ends. Hot air is blown into the parison, which expands to the shape of the mould, including relief details such as threads and surface decoration. Once the polymer solidifies, you eject the product by opening the split mould. Blow-moulded containers do not have to be symmetrical and can incorporate handles, screw threads and undercut features.

Figure 2.11: Fizzy drink bottles are blow-moulded

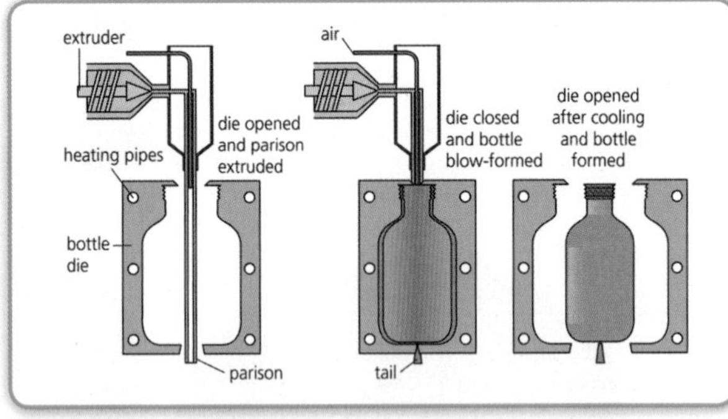

Figure 2.12: The blow-moulding process

Dome blowing

Dome blowing is the process of forming domes, spheres and oval shapes, usually out of acrylic. The sheet of acrylic is softened in an oven and transferred to a dome-blowing machine, where it is clamped under a circular ring. Air pressure is applied, which blows the material upwards and forms it into a perfect dome shape. Commercial dome blowing can produce perfect domes of a maximum diameter of two metres without distortion. This process is used to produce signage and point-of-sale displays.

Key terms are highlighted in the text, summarised at the end of each chapter, and a full definition given in the glossary at the end of the book to enable you to develop your understanding of Graphic Product terminology.

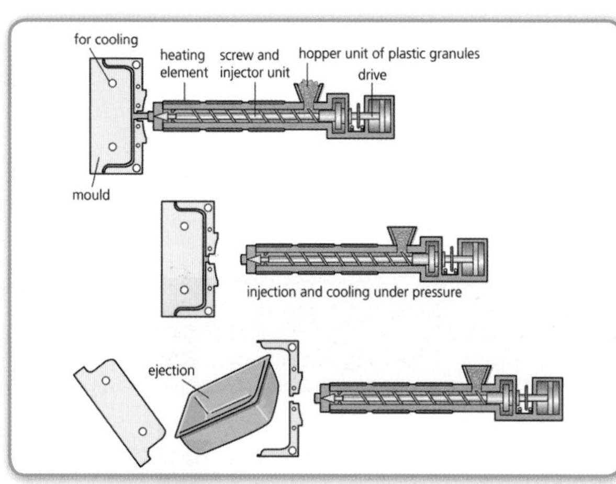

Figure 2.13: The injection-moulding process

Injection moulding

In the injection-moulding process, an expensive mould is injected with a liquid polymer, made by heating thermoplastic granules. Once the polymer cools and solidifies, the formed product is ejected. Injection moulding is suitable for complex shapes with holes, screw fittings and integral hinges.

Quality control (QC) inspection and testing

As many of these thermoforming processes are used for high batch or mass production, it is important that each product is identical in quality. Imperfect products have to be scrapped, and while many can be recycled, wastage still costs a company time and money. Companies will use several different types of machine to inspect batches of products at different stages of their production including laser measurement and ultrasonic testing.

- **Laser measurement** uses a series of lasers to measure the outside dimensions of the product as they move down a conveyor belt. This is a fast, accurate and repeatable process as lasers are computer-controlled and data is available immediately.
- **Ultrasonic testing** is used to check the wall thickness of a hollow product by sending high-frequency sound waves at the product as it moves down a conveyor belt. The waves will bounce back differently as they hit different materials and different thicknesses. The computer analyses the time it takes the sound waves to bounce back in order to calculate the wall thickness of the product accurately.

Support Activity

Make a list of products that are made using blow moulding and injection moulding. Collect images of these products for your revision notes so that you can give examples in an exam.

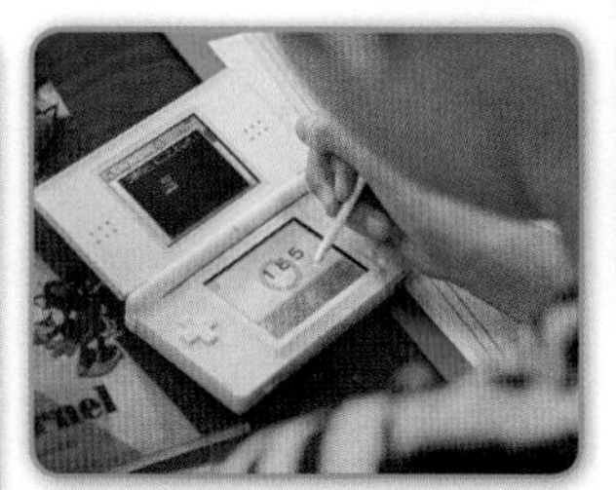

Figure 2.14 The casings for electrical products are usually injection-moulded

Engaging photos bring graphic products to life.

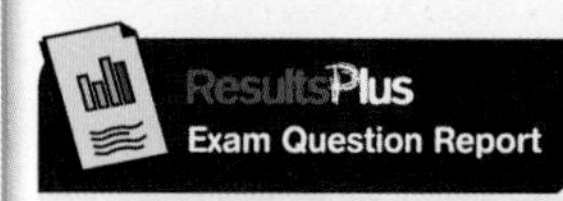

ResultsPlus
Exam Question Report

Explain two reasons why the injection-moulding process is suitable for mass production. (4 marks)

How students answered

Many students only stated one appropriate reason with no justification. Many of these students made simple statements such as 'cheaper', 'faster' and 'easier', which gained no marks at all.

36% 0-1marks

Most students could include two valid reasons why injection moulding is a suitable mass-production process. However, they only justified one reason in sufficient detail.

51% 2-3 marks

Some students correctly identified two appropriate reasons, fully justifying each. Remember: in an 'explain' question, you have to make a valid point and then go on to justify it. You cannot achieve full marks without any justification.

13% 4 marks

ResultsPlus features combine real exam performance data with examiner insight to give guidance on how to achieve better results.

Stretch Activity

Search the internet for animated diagrams of thermoforming processes so that you understand the process in detail.

Support and Stretch activities provide extra support to ensure understanding and opportunities to stretch your knowledge.

A dedicated suite of revision resources for **complete exam success.**

We've broken down the six stages of revision to ensure that you are prepared every step of the way.

Zone In How to get into the perfect 'zone' for your revision.

Planning Zone Tips and advice on how to effectively plan your revision.

Know Zone All the facts you need to know.

Chapter overview Outlines the key issue that the chapter examines. Keep this issue in mind as you work through the Know Zone pages.

Key terms Lists important graphic products terminology.

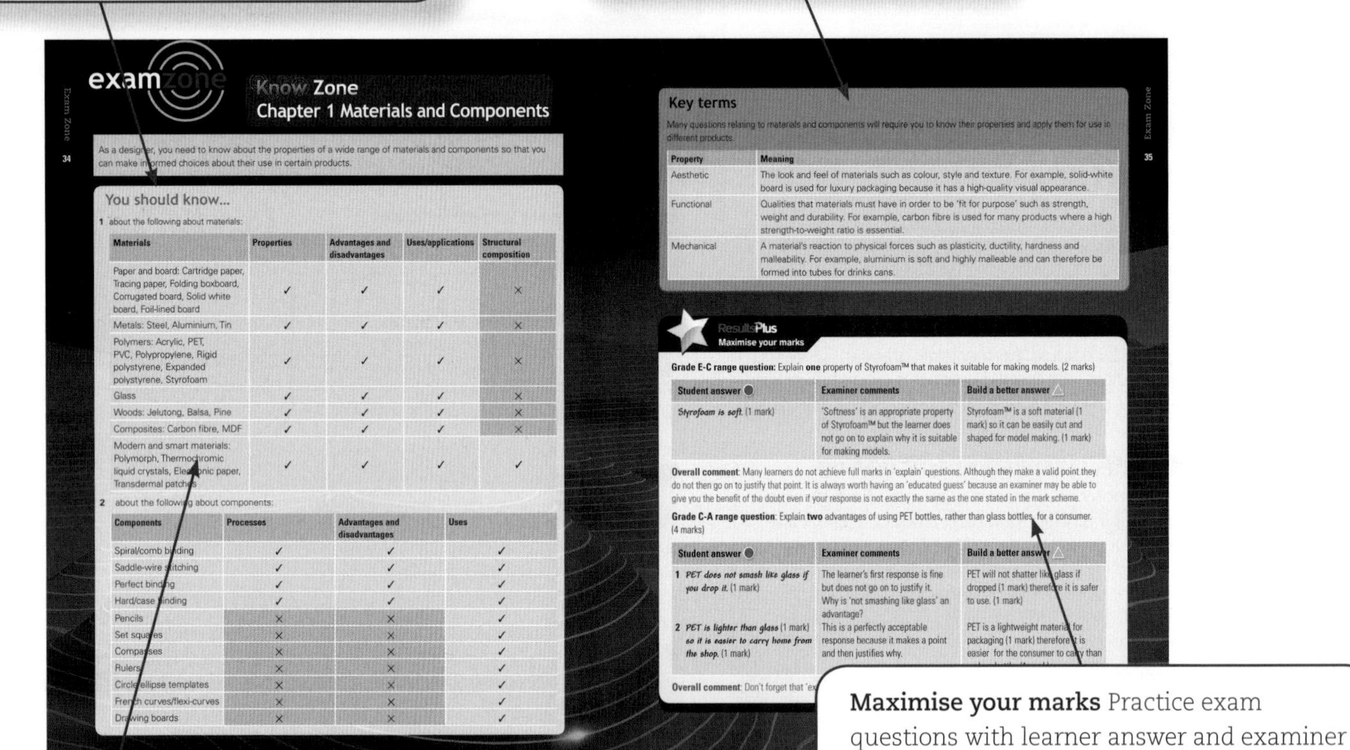

Know Zone
Chapter 1 Materials and Components

As a designer, you need to know about the properties of a wide range of materials and components so that you can make informed choices about their use in certain products.

You should know...

1 about the following about materials:

Materials	Properties	Advantages and disadvantages	Uses/applications	Structural composition
Paper and board: Cartridge paper, Tracing paper, Folding boxboard, Corrugated board, Solid white board, Foil-lined board	✓	✓	✓	✗
Metals: Steel, Aluminium, Tin	✓	✓	✓	✗
Polymers: Acrylic, PET, PVC, Polypropylene, Rigid polystyrene, Expanded polystyrene, Styrofoam	✓	✓	✓	✗
Glass	✓	✓	✓	✗
Woods: Jelutong, Balsa, Pine	✓	✓	✓	✗
Composites: Carbon fibre, MDF	✓	✓	✓	✗
Modern and smart materials: Polymorph, Thermochromic liquid crystals, Electronic paper, Transdermal patches	✓	✓	✓	✓

2 about the following about components:

Components	Processes	Advantages and disadvantages	Uses
Spiral/comb binding	✓	✓	✓
Saddle-wire stitching	✓	✓	✓
Perfect binding	✓	✓	✓
Hard/case binding	✓	✓	✓
Pencils	✗	✗	✓
Set squares	✗	✗	✓
Compasses	✗	✗	✓
Rulers	✗	✗	✓
Circle/ellipse templates	✗	✗	✓
French curves/flexi-curves	✗	✗	✓
Drawing boards	✗	✗	✓

Key terms

Many questions relating to materials and components will require you to know their properties and apply them for use in different products.

Property	Meaning
Aesthetic	The look and feel of materials such as colour, style and texture. For example, solid-white board is used for luxury packaging because it has a high-quality visual appearance.
Functional	Qualities that materials must have in order to be 'fit for purpose' such as strength, weight and durability. For example, carbon fibre is used for many products where a high strength-to-weight ratio is essential.
Mechanical	A material's reaction to physical forces such as plasticity, ductility, hardness and malleability. For example, aluminium is soft and highly malleable and can therefore be formed into tubes for drinks cans.

ResultsPlus
Maximise your marks

Grade E-C range question: Explain **one** property of Styrofoam™ that makes it suitable for making models. (2 marks)

Student answer	Examiner comments	Build a better answer
Styrofoam is soft. (1 mark)	'Softness' is an appropriate property of Styrofoam™ but the learner does not go on to explain why it is suitable for making models.	Styrofoam™ is a soft material (1 mark) so it can be easily cut and shaped for model making. (1 mark)

Overall comment: Many learners do not achieve full marks in 'explain' questions. Although they make a valid point they do not then go on to justify that point. It is always worth having an 'educated guess' because an examiner may be able to give you the benefit of the doubt even if your response is not exactly the same as the one stated in the mark scheme.

Grade C-A range question: Explain **two** advantages of using PET bottles, rather than glass bottles, for a consumer. (4 marks)

Student answer	Examiner comments	Build a better answer
1 *PET does not smash like glass if you drop it.* (1 mark)	The learner's first response is fine but does not go on to justify it. Why is 'not smashing like glass' an advantage?	PET will not shatter like glass if dropped (1 mark) therefore it is safer to use. (1 mark)
2 *PET is lighter than glass* (1 mark) *so it is easier to carry home from the shop.* (1 mark)	This is a perfectly acceptable response because it makes a point and then justifies why.	PET is a lightweight material for packaging (1 mark) therefore it is easier for the consumer to carry than

Overall comment: Don't forget that 'ex

Maximise your marks Practice exam questions with learner answer and examiner commentary (see next page).

You should know A check-yourself list of the concepts and facts that you should know before you sit the exam. Use this list to identify your strengths and weaknesses so you can plan your revision wisely.

Don't Panic Zone Last-minute revision tips for just before the exam.

Exam Zone An explanation of the assessment objectives, plus a chance to see what a real exam paper might look like.

Zone Out What do you do after your exam? This section contains information on how to get your results and answers to frequently asked questions on what to do next.

ResultsPlus

These features are based on the actual marks that learners have achieved in past exams. They are combined with expert advice and guidance from examiners to show you **how to achieve better results**.

There are four different types of ResultsPlus feature throughout this book.

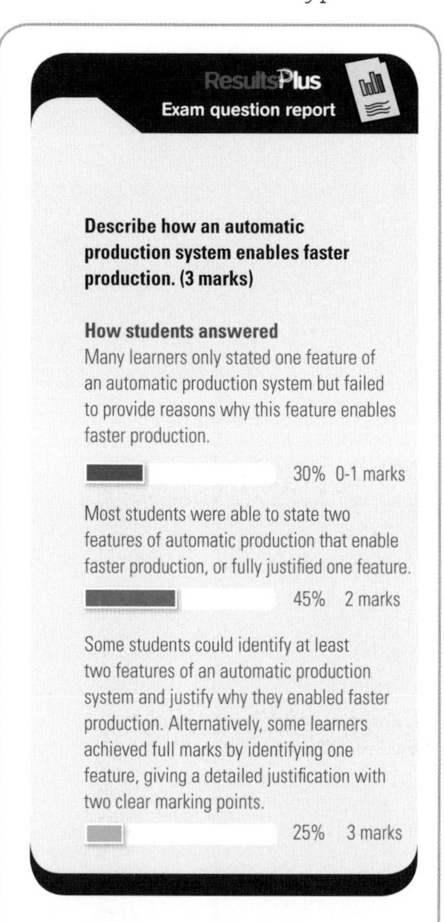

Describe how an automatic production system enables faster production. (3 marks)

How students answered

Many learners only stated one feature of an automatic production system but failed to provide reasons why this feature enables faster production.

30% 0-1 marks

Most students were able to state two features of automatic production that enable faster production, or fully justified one feature.

45% 2 marks

Some students could identify at least two features of an automatic production system and justify why they enabled faster production. Alternatively, some learners achieved full marks by identifying one feature, giving a detailed justification with two clear marking points.

25% 3 marks

Exam question report: These show previous exam questions with details about how well students answered them.

- Red shows the number of learners who scored low marks (less than 35% of the total marks)
- Orange shows the number of learners who did okay (scoring between 35% and 70% of the total marks)
- Green shows the number of learners who did well (scoring over 70% of the total marks).

They explain how learners could have achieved the top marks so that you can make sure that you answer these questions correctly in future.

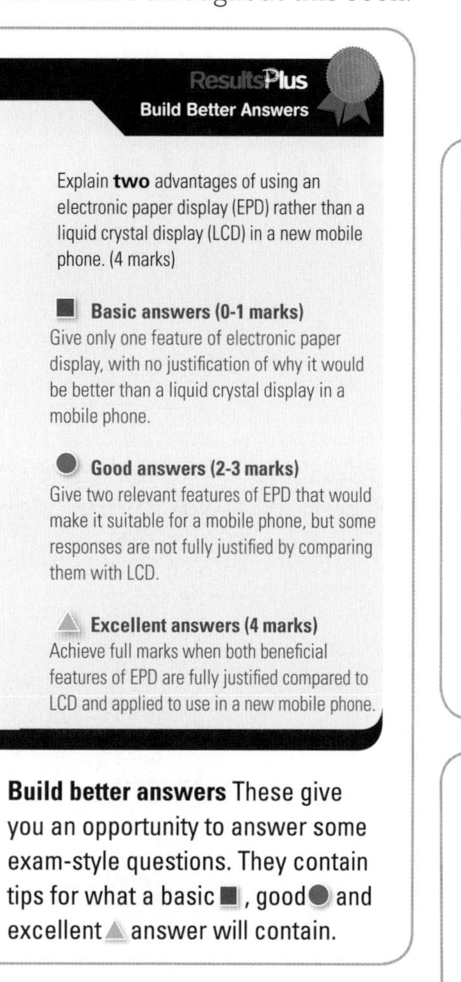

Explain **two** advantages of using an electronic paper display (EPD) rather than a liquid crystal display (LCD) in a new mobile phone. (4 marks)

■ **Basic answers (0-1 marks)**
Give only one feature of electronic paper display, with no justification of why it would be better than a liquid crystal display in a mobile phone.

● **Good answers (2-3 marks)**
Give two relevant features of EPD that would make it suitable for a mobile phone, but some responses are not fully justified by comparing them with LCD.

▲ **Excellent answers (4 marks)**
Achieve full marks when both beneficial features of EPD are fully justified compared to LCD and applied to use in a new mobile phone.

Build better answers These give you an opportunity to answer some exam-style questions. They contain tips for what a basic ■, good ● and excellent ▲ answer will contain.

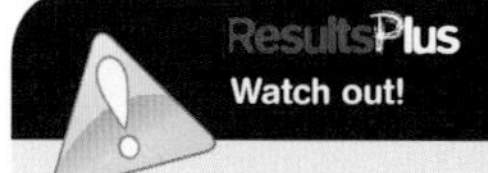

You may be asked to fill in blank spaces in a risk assessment table for a specific piece of equipment or process. You should familiarise yourself with safe working practices for a range of workshop equipment and processes.

Watch out! These warn you about common mistakes and misconceptions that examiners frequently see students make. Make sure that you don't repeat them! The ■, ● and ▲ symbols highlight the severity of the error.

Maximise your marks These are featured in the Know zone (see page 8) at the end of each chapter. They include an exam-style question with a student answer, examiner comments and an improved answer so that you can see how to build a better response. Follow the green triangles to create a full mark answer.

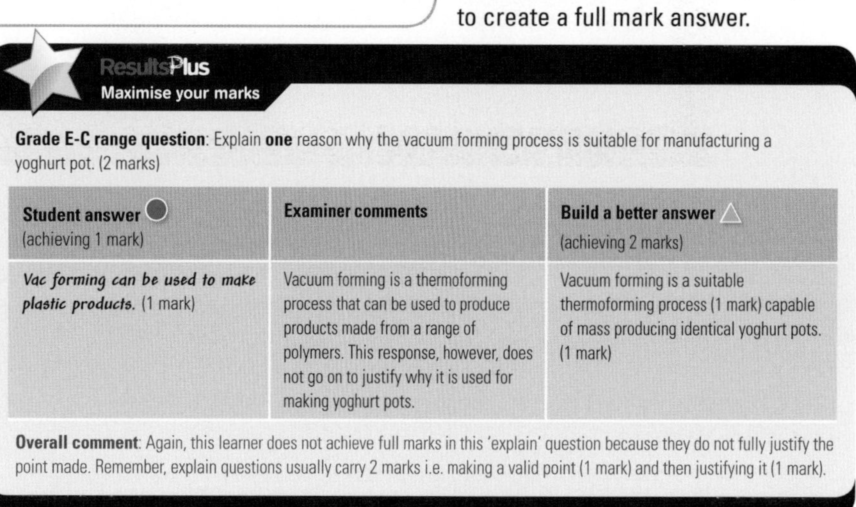

Grade E-C range question: Explain **one** reason why the vacuum forming process is suitable for manufacturing a yoghurt pot. (2 marks)

Student answer ● (achieving 1 mark)	**Examiner comments**	**Build a better answer** ▲ (achieving 2 marks)
Vac forming can be used to make plastic products. (1 mark)	Vacuum forming is a thermoforming process that can be used to produce products made from a range of polymers. This response, however, does not go on to justify why it is used for making yoghurt pots.	Vacuum forming is a suitable thermoforming process (1 mark) capable of mass producing identical yoghurt pots. (1 mark)

Overall comment: Again, this learner does not achieve full marks in this 'explain' question because they do not fully justify the point made. Remember, explain questions usually carry 2 marks i.e. making a valid point (1 mark) and then justifying it (1 mark).

Knowledge and understanding of Graphic Products

This unit requires you to demonstrate your knowledge and understanding of seven areas related to graphic products in a 1½ hour exam. The areas are covered in chapters.

Chapter 1: Materials and components You need to understand a wide range of materials and components to make informed choices about their use.

Chapter 2: Industrial and commercial processes You need to be able to design products that can be made, and that fit with modern processes, such as mass production.

Chapter 3: Analysing products This gives you important insights into meeting a design brief, commercial design and industrial manufacturing processes.

Chapter 4: Designing products Being creative and designing products is probably one of the reasons why you are studying Graphic Products.

Chapter 5: Technology The modern world relies on technology, which plays a key role in designs and creating modern graphic products.

Chapter 6: Sustainability Sustainable product design involves using energy and materials in a way that minimises the depletion of finite resources, waste production and pollution.

Chapter 7: Ethical design and manufacture Today's mass-consumer society creates a range of ethical issues that any designer needs to consider.

How much is it worth? Unit 2: Knowledge and Understanding of Graphic Products is worth **40%** of your overall GCSE Graphic Products course. This is an important exam, so familiarise yourself with its structure, take time to revise and practice your exam technique well.

SOLD
SOLD

Chapter 1 Materials and components

Paper and board

Objectives

- **Describe** the aesthetic, functional and mechanical properties of paper and board.
- **Explain** the reasons for selecting specific paper and board for graphic products and commercial packaging.
- **Understand** the composition of a packaging laminate.

Number of ply	Microns
2	200
3	230
4	280
6	360
8	500
10	580
12	750

Table 1.1: Common thicknesses of paper and board

Weight and size

Paper is classified by weight in gsm (grams per square metre), with 80 gsm being the weight of average copier paper. Board is measured in microns (micrometers, or 1000ths of a millimetre). The thickness of a board can be gauged by the number of ply (layers) it is made up of. A paper usually becomes reclassified as a board when it is heavier than 220 gsm, and more often than not is made from more than one ply.

Paper and board are most commonly available in metric 'A' sizes (A5 through to A0). However, 'B' sizes and old imperial measurements are also widely used.

Paper

The choice of paper is essential to how printed graphic products are presented. Choosing the most appropriate paper for the end product involves a combination of personal preference and discussion with the client. In general, the correct choice of paper must satisfy:

- the design requirements of the client's brief, e.g. durability, surface finish, colour, texture
- the demands of the printing process or surface decoration, e.g. whether the printing inks used for offset lithography will provide a quality finish on the paper
- any economic considerations, e.g. scale of production.

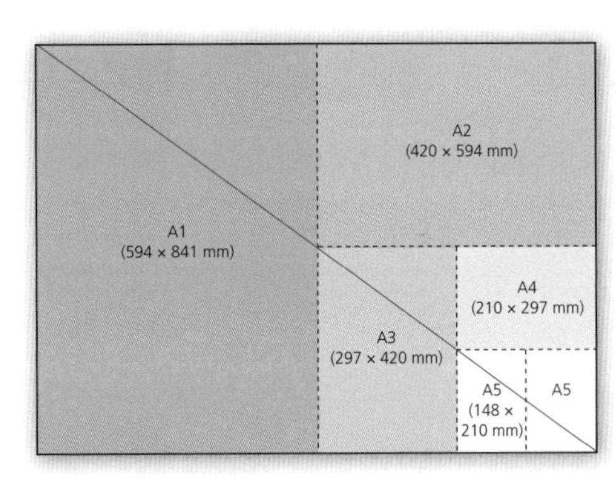

Figure 1.1: Common 'A' sizes of paper and board

Type	Weight	Description	Uses	Properties	Cost
cartridge paper	120-150 gsm	• creamy-white paper • smooth surface with a slight texture	• good general purpose drawing paper • heavier weights can be used with paints	• completely opaque • accepts most drawing media	more expensive than copier paper
tracing paper	60/90 gsm	• thin, transparent paper with a smooth surface • pale grey in appearance	• same as layout paper • heavier weight preferred by draughtspeople	allows tracing through on to another sheet in order to develop design ideas	heavier weight can be quite expensive

Table 1.2: Paper types

Board

Cartonboard is the generic name given to board used widely in the retail packaging industry, where specific properties are required. These boards must be suitable for high-quality, high-speed printing and for cutting, creasing and gluing using high-speed automated packaging equipment.

The main advantages of using cartonboard for commercial packaging include:

- excellent print quality on most boards
- excellent protection in structural packaging nets
- relatively inexpensive to produce and process
- it can be recycled readily.

Table 1.3 shows you why different types of cartonboard are used for packaging different products.

Type	Description	Properties	Uses	Cost
folding boxboard	usually consists of a bleached virgin pulp top surface, unbleached pulp middle layers and a bleached pulp inside layer	• excellent for scoring, bending and creasing without splitting • excellent printing surface	most food packaging and all general carton applications	relatively inexpensive
corrugated board	constructed from a fluted paper layer sandwiched between two paper liners	• excellent impact resistance • excellent strength for weight • recyclable	protective packaging for fragile goods; the most commonly used box-making material	relatively inexpensive
solid white board	made entirely from pure, bleached wood pulp	• very strong and rigid • excellent printing surface	packaging for frozen foods, ice-cream, pharmaceuticals and cosmetics	expensive
foil-lined board	board with a laminated foil coating (the coating can be used on all of the boards above)	• very strong visual impact • foil provides excellent barrier against moisture	cosmetic cartons, pre-packed food packages	expensive

Table 1.3: Cartonboards for commercial packaging

Apply it!

You will be able to use a range of paper and boards in Unit 1: Creative Design and Make Activities.

Folding boxboard and solid white board to make your final packaging nets out of are readily available from school suppliers.

Explain **two** benefits of using corrugated board for packaging electrical products.(4 marks)

Basic answers (0-1 marks)
Give only one generic property of corrugated board, with no justification of why it is used for the packaging.

Good answers (2-3 marks)
Give two specific properties of corrugated board that make it suitable for the packaging, but some responses are not fully justified.

Excellent answers (4 marks)
Achieve full marks when both properties of corrugated board are fully justified and applied to its use in the packaging of electrical products.

Packaging laminate

A polyethylene, aluminium foil and paperboard packaging laminate combines the characteristics of all three materials into a single package. Developed by Tetra Pak in the 1950s, this innovation has changed the way food is packaged and distributed around the world.

Tetra Pak's strength is in the area of hygienic processing and packaging of drinks. When packaging laminate is combined with ultra-high-temperature processing (UHT), products can be packaged and stored under room temperature conditions for up to a year. This allows for perishable goods to be saved and distributed over greater distances.

Figure 1.2: Tetra Pak cartons make use of a packaging laminate

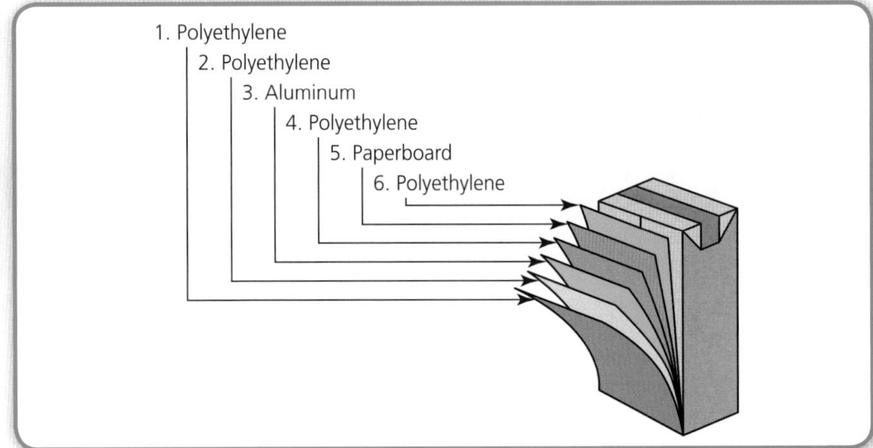

Figure 1.3: Cross-section of a packaging laminate

Layer	Material	Reason for use
1	polyethylene	external layer provides protection against outside moisture
2	polyethylene	helps the internal layer bond with the aluminium foil
3	aluminium foil	provides a barrier against the harmful effects of air and light
4	polyethylene	helps the paperboard bond with the aluminium foil
5	paperboard	• provides stiffness and shape • excellent print surface for total graphic coverage
6	polyethylene	internal layer, in contact with the drink, seals in the liquid

Table 1.4: Materials in the layers of a packaging laminate

ResultsPlus
Build Better Answers

Describe how a packaging laminate, such as a Tetra Pak carton, keeps the liquid inside fresh. (5 marks)

■ **Basic answers (1-2 marks)**
Refer to one material only, with limited justification of its use in the packaging laminate.

● **Good answers (3-4 marks)**
Refer to specific properties of aluminium and polyethylene that make them suitable for keeping the liquid fresh, but some responses are not fully justified.

▲ **Excellent answers (5 marks)**
Achieve full marks when the properties of aluminium and polyethylene are fully justified and applied to use in the packaging laminate.

Environmental issues with paper and board

Paper and board are produced primarily from hard and softwoods. Wood is made up of cellulose fibres bound together by lignin. To produce paper, these fibres must be separated to form a mass of individual fibres called wood pulp. Softwood fibres are longer, offering greater strength; hardwood fibres are shorter, offering a smoother, opaque finish. Wood pulp is produced by one of three methods: mechanical, chemical or waste pulping.

Toxic pollution

Many toxic chemicals are used in paper making, especially toxic solvents and chlorine compounds used to bleach and remove the lignin from pulp. Many pulp and paper mills have made reductions in chemical use over the years, but these benefits may be countered by increased production.

Air pollution

Pulp and paper mills are major sources of air pollutants, such as carbon dioxide. These gases contribute to the depletion of the ozone layer, global warming, acid rain and respiratory problems in the local area.

Water and energy consumption

Paper making is energy-intensive, using large amounts of electricity. It also uses a great deal of water from local supplies, which can cause problems such as increased sedimentation, increased water temperature, a reduction in the number of wildlife species in local habitats, possible concentration of toxic material and lowering of water tables.

Solid waste

Paper fibres can only be recycled a limited number of times before they become too short or weak to make high-quality paper. The broken, low-quality fibres are separated out to become waste sludge filling large areas of landfill space each year. Some companies may burn their sludge in incinerators.

Deforestation

Worldwide, enormous areas of virgin forest are being felled for paper pulp production, contributing to the destruction of the world's forests or deforestation. Many UK paper mills import their pulp, some of which may come from endangered forests.

Possible solutions

Mechanical pulping is arguably the least environmentally damaging process for producing virgin wood pulp, and has the advantage of producing a high pulp yield compared to other processes. Newsprint quality, mechanically-pulped paper contains few additives, can be recycled up to four times and is adequate for many purposes for which chemically pulped products are currently used.

One solution would be greater use of recycled fibre pulp rather than virgin pulp. The UK currently imports large quantities of virgin pulp, so needs to expand its recycling industry as an alternative.

Using alternative fibres, such as hemp and straw, can reduce pressure on forests, and have great potential for being produced in the UK.

Support Activity

1. Why would a 'virgin' pulp be used to make the paper for a glossy fashion magazine? Explain your answer.
2. Why is recycled pulp used for newpapers? Explain your answer.

Stretch Activity

Using the internet and the library, investigate commercially available packaging nets. Find out what products you could use to form your own boxes for design tasks.

Metals

Objectives

- **Describe** the aesthetic, functional and mechanical properties of steel, aluminium and tin.
- **Explain** the reasons for selecting specific metals for graphic products and commercial packaging.
- **Understand** the environmental implications and advantages and disadvantages of using metals.

Figure 1.4: Drinks cans can be made from either aluminium or steel

Figure 1.5: The open-cast mining of bauxite destroys vast areas of land

Ferrous and non-ferrous metals

Metals can be divided into two main groups:

- ferrous metals, which contain mainly ferrite or iron (e.g. steel), almost all of which are magnetic. This group also includes metals with small additions of other substances, such as carbon steels
- non-ferrous metals, which contain no iron (e.g. aluminium and tin) – these are not magnetic.

Steel

When steel is rolled into thin sheets, it is a lightweight material that can be easily formed into a range of shapes. Steel is used to make a range of food cans and most aerosol cans.

Aluminium

Aluminium is a strong but lightweight metal that is easily formed into a variety of shapes. It is a non-ferrous metal so it does not rust, which makes it perfect for making a range of containers.

Tin

Tin is usually used in commercial packaging in the form of tinplate, which consists of cold-rolled steel sheet coated with a thin layer of tin to prevent corrosion. This is why cans are often called 'tin cans'.

Environmental issues

Steel is produced from iron ore, which is widely found and mined. It takes a lot of energy to turn iron ore into steel as it has to be heated in a huge furnace to very high temperatures, to separate it from the other materials present. However, steel can be easily recycled, and it takes about 75 per cent less energy to recycle steel than to make it from iron ore. Steel is a ferrous metal, and is magnetic, making it easy to sort using powerful electro-magnets at recycling plants, which separate the steel from other metals. Currently, around 51 per cent of all steel packaging is recycled.

Aluminium is a pure, naturally occurring metal element that is mined from beneath the land and sea. It is the most plentiful metal element in the earth's crust and is produced from the ore bauxite. The extraction of alumina from the bauxite ore, and subsequent production of aluminium from alumina, uses up a lot of energy. However, aluminium is easily recycled as it can be melted down and reused over and over again without being spoiled. Recycling aluminium is much more sustainable than producing it from alumina, because it saves considerable amounts of energy. Recycling one aluminium can saves enough energy to run a TV for three hours. If all the aluminium cans in Britain were recycled, there would be about 14 million fewer dustbins of waste every year. Currently, just 34 per cent of aluminium drinks cans are recycled.

Tin is not a naturally occurring metal and must be extracted from the ore cassiterite using a coal-fired reverberatory furnace. This process uses a lot of energy.

Both aluminium and tin are extracted using open-cast mining on land or dredging of the sea bed. These techniques destroy large amounts of land, which has to be expertly managed if it is to be reinhabited after mining has finished. Dredging the seabed can destroy sealife and aquatic habitats.

Metals in graphic products

Steel and aluminium have a wide range of uses, but their biggest use in graphic products is for exterior signage. Effective exterior signage can be crucial in a highly competitive business world: it is often the key to maximising a location and helping a company to 'stand out from the crowd'. Metal can be fabricated into relatively complex shapes to provide tailor-made solutions for shop frontages, or made into standard components such as pavement signs.

Aluminium is ideal for exterior use because it does not corrode. When steel is used, a surface finish is required such as a clear lacquer, or powder coating, which is available in a range of colours (think of climbing frames in the park). Stainless steel could be used for some products because it has added carbon for strength and chromium for resistance to wear and corrosion. Stainless steel has a high-quality surface finish, which does not require additional finishes.

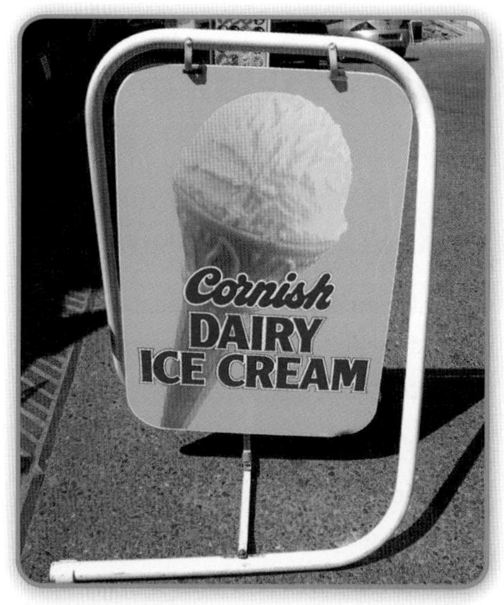

Figure 1.6: A swinging pavement sign made from aluminium

Figure 1.7: Metal containers provide visual impact for luxury products

Metals in commercial packaging

Metals are used in a wide range of commercial packaging applications, from the standard components of drinks cans and bottle tops to special promotional containers that suggest quality. There are several advantages in using metals for commercial packaging, including:

- added security, as sealed cans cannot be tampered with without obvious visible signs
- containers can be made in a variety of standard sizes and shapes (e.g. drinks cans) and custom-made styles (e.g. chocolate and biscuit tins)
- containers can be embossed or de-bossed to provide surface textures and visual appeal
- metals can be directly printed onto, or a paper label can be added for total graphic coverage, providing an effective point-of-sale display.

Metal appeal

Metal packaging suggests quality, and manufacturers will often use it for special promotions. For example, during the Christmas period many brands of chocolate and biscuits will be available in large and highly decorative metal tins. Some biscuit tins dating from the nineteenth century have even become collectors' items. Metal tins also have the advantage of being re-usable and not simply thrown away after the contents have been eaten!

Support Activity

Look at the picture of the swinging pavement sign above.

1. Why is aluminium a suitable material for this graphic product? Explain your answer.
2. What process was used to print the design onto the aluminium sign?

Stretch Activity

1. Research the techniques of 'embossing' and 'de-bossing' for adding visual appeal to metal tins.
2. Find out how aluminium drinks cans are made using 'cold forming'.

Polymers

Objectives

- **Describe** the aesthetic, functional and mechanical properties of polymers.
- **Explain** the reasons for selecting specific polymers for graphic products and commercial packaging.
- **Understand** the environmental implications and the advantages and disadvantages of using polymers.

Polymers or 'plastics' are a relatively modern material. They were first produced in the early twentieth century; their use grew rapidly during the second half of the century, and continues to grow today. Polymers have provided alternatives to many packaging requirements previously carried out by metal, glass and boards, and have often completely replaced these materials.

Thermoplastics

Thermoplastics are polymers that, once heated, can be formed into a variety of interesting shapes using different forming techniques such as blow moulding, vacuum forming and injection moulding (see pages 44–46 for more on these techniques). Once the polymer has cooled down, the shape then remains permanent. The same thermoplastic can be heated, softened, shaped and cooled many times over, which means these materials are ideal for recycling.

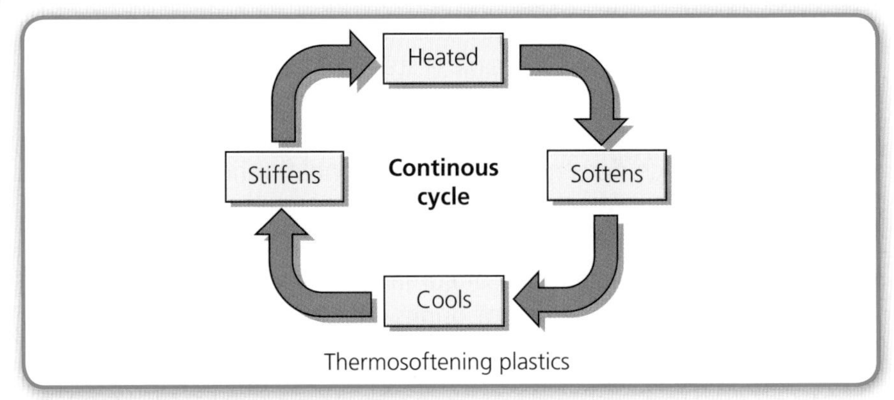

Figure 1.8: Thermoplastic heating cycle

Thermoplastics are used in a wide range of graphic products, from the Styrofoam™ models that you make in the classroom, to the shop signage you see on the high street and the injection-moulded casings of many electrical products, such as the games console in Figure 1.9.

Polymers in electrical products

Many electrical products make use of polymers to form their protective casings. Polymers have a number of advantages over metal casings:

- they are less expensive to produce in large quantities and thermoforming processes, such as injection moulding, can produce extremely intricate shapes
- they are non-conductive – they do not conduct electricity – so products are safe to handle
- they have excellent heat-resistant properties, which is an advantage where electronic components can become quite hot with prolonged use (just think about your games console when you have been playing a game for a long time).

Figure 1.9: The casings of electrical products are often made from polymers

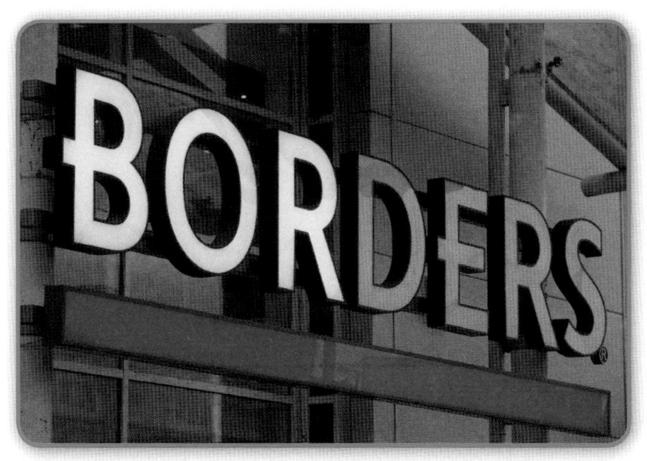

Figure 1.10: Acrylic is used to great effect in shop signage

Acrylic

Acrylic is usually cast into sheets but is also available in rods and tubes. It is self-finishing, so it does not need to be painted. Acrylic has a wide range of uses within graphic products, from making models to shop signage.

When making a model, acrylic is a versatile sheet material as it can be easily cut and bent using a strip heater. Thinner sheets of acrylic can also be vacuum-formed just like polystyrene or PVC. Acrylic is available in a wide range of colours, and has an excellent surface finish, which adds a quality feel to any model.

In shop signage, a higher quality acrylic is used which must satisfy several high performance requirements. It has to be able to:

- withstand extreme weather conditions, such as heat in the summer and cold in winter
- be chemical-resistant to pollution and detergents
- be durable by resisting long-term stresses, such as being. outdoors for prolonged periods
- be easy to fabricate and to make relatively complicated shapes out of
- have excellent aesthetic properties, to attract a customer's attention and give a high-quality look for the business.

Styrofoam™

Styrofoam™ is a polystyrene foam manufactured for the construction industry for insulating buildings. It also makes an excellent modelling material. The process of making the foam results in a material with uniformly small, closed cells, which gives it all of the properties that a modelling material should have, including:

- great rigidity and high compressive strength so it cannot be bent out of shape easily
- easy to cut and shape with a range of hand tools
- sheets can be glued together using PVA to form larger blocks
- a smooth surface finish when sanded
- can be painted to give a good quality finish if desired.

Apply it!

You will be able to use both Styrofoam™ and acrylic in Unit 1: Creative Design and Make Activities.

Styrofoam™ could be used at the development stage of your design activity to test and refine your ideas. For example, you could find out more about ergonomic factors by producing rough 3D models that can be physically handled.

Acrylic is an extremely useful material to use in the manufacture of your final product.

Support Activity

1 What are the advantages of using acrylic signage for a shop instead of metal or wooden signage?

2 Why are metal casings sometimes used for electrical products in preference to injection-moulded casings made from polymers?

Stretch Activity

Study a simple product such as an MP3 player. Take accurate measurements and make an accurate Styrofoam™ model of it. You could make a full-scale (1:1) model or use a suitable scale, such as half size (1:2), if appropriate.

ResultsPlus
Build Better Answers

Explain **two** reasons why polystyrene is used for the casing of many electrical products. (4 marks)

Basic answers (0-1 marks)
Give only one generic property of polystyrene, with no justification of why it is used for the casing.

Good answers (2-3 marks)
Give two specific properties of polystyrene that make it suitable for the casing, but some responses are not fully justified.

Excellent answers (4 marks)
Achieve full marks when both properties of polystyrene are fully justified and applied to its use in the casings of electrical products.

Figure 1.11: Bottles made from polymers can be collected, sorted and recycled

Polymers in commercial packaging

Polymers are widely used in commercial packaging because they are:

- lightweight and versatile
- strong, tough, rigid, durable, impact and water-resistant
- easily formed and moulded
- easy to print on
- low-cost
- recyclable.

Polymers can be identified by a code, usually stamped onto the base of the package or printed onto the label. An internationally recognised coding system means that you can easily identify polymers and sort them for recycling. Each polymer has its own useful properties, making it suitable for use in different areas of the packaging industry (see Table 1.5).

Expanded polystyrene

Expanded polystyrene is used in fast-food packaging because it is:

- hygienic – disposable cups and plates mean germs and bacteria are simply thrown away with the rubbish instead of multiplying in a chipped coffee mug, for instance.
- strong, yet lightweight – protects against moisture and keeps its strength. Containers and lids close tightly and prevent any leakage of the contents. It can be moulded into a variety of structural packages which complement its excellent cushioning properties in protecting the contents of the package.
- efficient – provides excellent insulation. Therefore, hot food can be kept warm for longer periods. It also means that the package does not become so hot that it cannot be held in the hand.
- economical – products are generally cheaper to buy than disposable paper products and much cheaper than reusable service ware (e.g. china). This is because only about five per cent of the foam package is actually plastic – the rest is simply air!
- convenient – with today's busy lifestyles people want food to be available instantly, and polystyrene is an economical way of serving people with their fast food.

Expanded polystyrene is also used in the protective packaging of many delicate products, such as electrical products; shape-moulded, it fits snugly around products which are placed inside corrugated board boxes.

Disadvantages of polymers

Environmental concerns

The main disadvantage for the widespread use of polymers centres on concerns about how sustainable they are. They are made from oil, which is a finite resource that requires a lot of energy to process, producing high levels of pollution. Manufacturing techniques for polymer-based packaging, such as blow moulding, also consume a lot of energy. Then comes the problem of disposal. Polymers are durable and degrade slowly, which is a problem for landfill sites.

Polymer	Properties	Uses
1 PET polyethylene terephthalate	• excellent barrier against atmospheric gases • prevents gas from escaping package • does not flavour the contents • sparkling 'crystal clear' appearance • very tough • lightweight – low density	• bottles for carbonated (fizzy) drinks • packaging for highly flavoured food • microwaveable food trays
3 PVC polyvinyl chloride	• weather resistant – does not rot • chemical resistant – does not corrode • protects products from moisture and gases while holding in preserving gases • strong, good abrasive resistance and tough • can be manufactured either rigid or flexible	packaging for toiletries, pharmaceutical products, food and confectionery, water and fruit juices
5 PP polypropylene	• lightweight • rigid • excellent chemical resistance • versatile – can be stiffer than polyethylene or very flexible • low moisture absorption • good impact resistance	• food packaging – yoghurt and margarine pots, sweet and snack wrappers • used for laminating paper and board
6 PS polystyrene	**rigid polystyrene:** • good impact resistance • rigid • lightweight • low water absorption **expanded polystyrene:** • excellent impact resistance • very good heat insulation • durable • lightweight • low water absorption	**rigid polystyrene:** food packaging such as yoghurt pots, CD jewel cases **expanded polystyrene:** egg cartons, fruit, vegetable and meat trays, cups, etc.; packing for electrical and fragile products

Table 1.5: Polymers commonly used in commercial packaging

Support Activity

Fast-food packaging debate

We know that there are advantages to having our food served in disposable containers, but what are the environmental issues in using them?

Divide into two teams:

- Team 1 – for disposable fast-food packaging
- Team 2 – against disposable fast-food packaging.

Have a debate and see which team wins!

Stretch Activity

Use the internet to research the system for collecting, sorting and recycling polymers. Your local council website would be a good place to start.

Glass

Objectives

- **Describe** the aesthetic, functional and mechanical properties of glass.
- **Explain** the reasons for selecting glass for commercial packaging.
- **Understand** the automatic processes for mass-producing glass containers.

ResultsPlus
Build Better Answers

Explain **two** advantages, to the consumer, of a bottle being made out of PET instead of glass. (4 marks)

Basic answers (0-1 marks)
Give only one generic property of PET, with no justification of why it is better than glass for a bottle.

Good answers (2-3 marks)
Give two specific properties of PET but only one reason why these are better properties than glass for a bottle for the consumer.

Excellent answers (4 marks)
Achieve full marks when two properties of PET are fully justified as advantages for the consumer over glass for a bottle.

Glass in commercial packaging

Glass is one of the earliest materials used for containing food and drink, and continues to be used for a range of packages, from jam jars to expensive perfume bottles. Glass has many properties that make it ideal for use as a packaging material. It is:

- cost-effective when mass-produced
- resistant to mechanical shock, as it is strong when annealed
- transparent, offering excellent product visibility
- an inert material, so it is safe and will not react with the contents
- relatively lightweight – today's glass containers are more than 40 per cent lighter than they were 20 years ago
- reusable and recyclable.

In addition:

- it does not deteriorate, corrode, stain or fade, so the contents remain fresh
- its contents can be preserved through high-temperature processing, such as sterilisation of milk
- it can be sealed air-tight, to protect the contents from contamination and bacteria.

Glass and the environment

- Glass packaging is 100 per cent recyclable – it can be recycled endlessly, with no loss in quality or purity, so recovered glass containers can be made directly into new glass bottles.
- Recycling glass saves raw materials – for every ton of glass recycled, over a ton of natural resources is conserved.
- Recycling lessens the demand for energy – energy costs drop when recovered glass is used in conjunction with virgin raw materials.
- Recycling cuts CO_2 emissions – for every six tons of recycled container glass used, one ton less of carbon dioxide is created.
- No processing by-products – glass recycling is a closed-loop system, creating no additional waste or by-products.

Colour

In its purest form, glass has a greenish tint. By adding chemicals in varying quantities to the raw mixture of sand, soda and limestone different colours of glass can be produced. This is useful when designing glass containers for a specific product. For example, beer is usually packaged in green or brown glass to protect it from direct sunlight.

Glass versus polymers

Glass has one major advantage over polymers: it has a more sophisticated image. For example, using a plastic beaker for vintage champagne would be inappropriate; glass, on the other hand, can give the product a look of quality. The greatest disadvantage with glass is that it can shatter.

Mass-producing glass containers

Note: These processes will not be tested in your exam but you may need them to analyse products made of glass, such as bottles.

1. Melting

The basic raw materials for making glass are quartz sand, limestone and soda ash, plus some oxides that act as catalysts to melt the glass. These are automatically mixed and fed into a furnace where they are heated at up to 1600°C to produce molten glass.

2. Shaping or forming

Molten glass is fed into machines where it is automatically blown using either the blow-and-blow process for bottles, or the press-and-blow process for wide-mouthed containers, such as jars.

3. Annealing

The hot glass containers undergo a slow and controlled cooling process to relieve internal stresses and make the glass stronger. Glass that has not been annealed is liable to crack or shatter easily when subjected to small changes in temperature or mechanical shock, such as being dropped.

4. Quality control inspection

Once cooled, the glass containers are channelled through inspection stations and are checked for dimensional accuracy, and body and neck quality. Inspection can be manual, semi-automatic or automatic.

5. Packing

After the inspection stations, the glass containers are put on pallets and protected with shrink-wrap, ensuring safe delivery to customers.

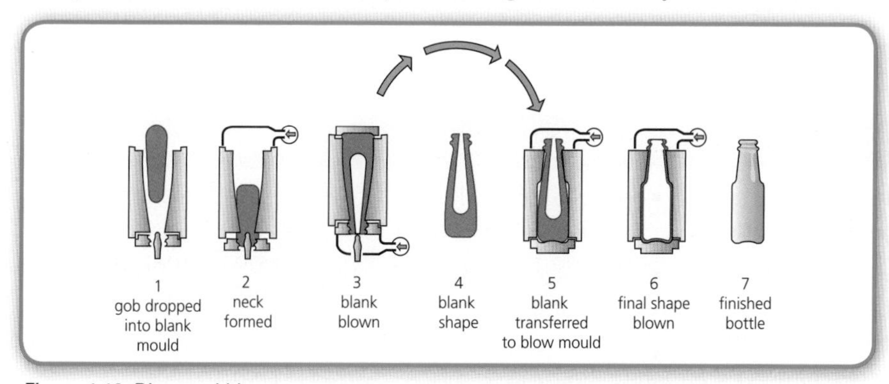

Figure 1.13: Blow and blow process

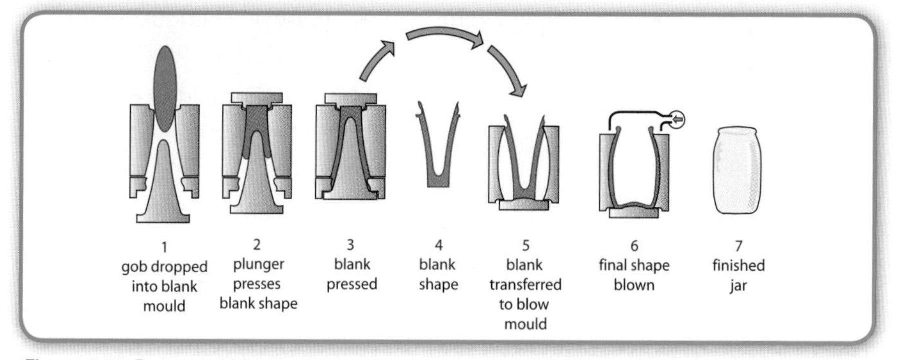

Figure 1.14: Press and blow process

Figure 1.12: The instantly recognisable Coca-Cola 'Contour' glass bottle

Support Activity

Why is glass used to package:

- perfumes and aftershave?
- more expensive brands of sparkling mineral water?

Explain your answers.

Stretch Activity

Use the internet to find out how manufacturers print designs directly onto glass.

Woods

Objectives

- **Describe** the aesthetic, functional and mechanical properties of hardwoods and softwoods.
- **Explain** the reasons for selecting specific woods for modelling and prototyping.
- **Understand** the environmental implications and advantages and disadvantages of using woods.

Figure 1.15: Balsa is often used for model aircraft because it is easy to shape and lightweight

Woods for modelling and prototyping

Wood can be extremely useful for producing a range of graphic products, from interior design and architectural models to product prototypes. It is available in a variety of shapes, sizes and thicknesses, and in a range of types that each has its own useful properties.

Hardwoods

Hardwoods are produced from broad-leaved trees whose seeds are enclosed; examples include oak, mahogany, beech, ash and elm. Hardwood trees commonly grow in warmer climates, such as Africa and South America, and take about 100 years to reach maturity. They are usually tough and strong and, because of their close grain, provide highly decorative surface finishes. Because of their age and location, many hardwoods are expensive to buy and so they are often reserved for high-quality products. The exceptions are jelutong and balsa, which have been used for modelling and pattern making for many years, as they are relatively inexpensive and easy to shape.

Softwoods

Softwoods are produced from cone-bearing trees (conifers) with needle-like leaves; examples include Scots Pine (red deal), Parana Pine and White Pine. As softwoods grow more quickly than hardwoods (taking around 30 years to mature), they can be forested and replanted, which means they are in abundance and so are cheaper to buy. Softwoods are also easier to work with and are lightweight, which makes them more suitable for modelling applications.

Wood	Description	Uses	Advantages	Disadvantages
jelutong	• straight-grained with fine, even texture • creamy-white colour	• model and pattern making • excellent wood to carve and sculpt due to its softness	• works easily with both hand and power tools • glues, screws, and nails without difficulty • stains, paints, and varnishes fairly well	• quite brittle and weak • low resistance to decay • cannot be easily steam-bent
balsa	• straight-grained and 'spongy' • pale beige to pinkish colour	• softest and lightest commercial hardwood • used for buoyancy aids and model-making	• very soft and light • quite strong for its weight • quite stable in use • possibly the easiest wood to cut, shape and sand • extremely buoyant (floats very well in water)	• low in strength, stiffness and shock resistance • finishes fairly well but porous composition soaks up finish • not suitable for steam-bending

Table 1.6: Hardwoods used in modelling and prototyping

Wood	Description	Advantages	Disadvantages	Uses
Parana Pine	• attractive with straight grain and very close density • honey colour	• hard, straight grain (often knot-free) • fairly strong, durable and easy to work • smooth finish • glues without difficulty and holds nails and screws quite well	• low stiffness, shock resistance and decay resistance • can distort significantly if not seasoned properly	furniture, turned items, general construction, plywood veneers and woodpulp for papermaking
White Pine	• straight-grained with a fine, uniform texture • creamy white, pale yellow or light brown • yellows with age	• works very easily with most machine or hand tools • carves quite well • holds nails and screws well without the need to pre-drill • glues, paints and varnishes well	• does not turn well • needs sealer to prevent blotches when staining • quite weak • low decay resistance, shock resistance • not good for steam-bending	carvings and sculpture, toys, novelties, musical instrument, boxes, match sticks, veneer, dowels and patterns

Table 1.7: Softwoods used in modelling and prototyping

Forest Stewardship Council

The Forest Stewardship Council (FSC) is an international, non-governmental organisation dedicated to promoting responsible management of the world's forests.

According to the FSC, every year, on every continent, an area half the size of the UK is cleared of natural forests. These forests are irreplaceable and their loss can have many economic, social and environmental impacts. Millions of indigenous people (including some of the world's poorest), types of plant and animal species are wholly dependent on forests for survival.

The FSC operates a global forest certification system that allows consumers to identify, purchase and use timber products produced from well-managed forests that do not contribute to the destruction of the world's forests.

Figure 1.17: Softwoods are used to make children's toys such as this jigsaw puzzle

Figure 1.16: The FSC logo is used on labels to indicate whether products are certified under the FSC system

Support Activity

Why are softwoods used for making children's toys, such as the jigsaw puzzle above, instead of hardwoods? Explain your answer.

Stretch Activity

Why are the rainforests of South America under threat? Why are they important?

Composites

Objectives

- **Describe** the aesthetic, functional and mechanical properties of carbon fibre and medium density fibreboard (MDF).
- **Explain** the reasons for selecting carbon fibre and MDF for a range of graphic products.
- **Understand** the health and safety issues when working with carbon fibre and MDF.

What is a composite material?

When two or more materials are combined by bonding, a composite material is formed. The resulting material has improved mechanical, functional and aesthetic properties and, with most composites, it will have an excellent strength-to-weight ratio.

Carbon fibre

Carbon fibre is a material consisting of extremely thin fibres of carbon. Several thousand carbon fibres are twisted together to form a yarn, which is then woven into a fabric. Carbon fibre has many different weave patterns. When combined with a polymer resin, it can be moulded to form a composite material called carbon fibre reinforced plastic. Depending on the direction of the fibres, the composite can be stronger in a certain direction or equally strong in all directions. The complex nature of carbon fibre makes it very difficult to break.

Carbon fibres are incredibly strong and so are ideal for high-performance structural uses such as aircraft frames, quality sports equipment and Formula 1 racing car manufacture. Carbon composites have excellent mechanical properties where high strength and low weight are needed, and will easily outperform any metal alternative. For example, carbon fibre has more than four times the tensile strength of the best steel alloys, at just a quarter of the weight! The main disadvantage of carbon fibre at present is its high cost, as it is still a specialised material that has not yet been used in mass-produced products.

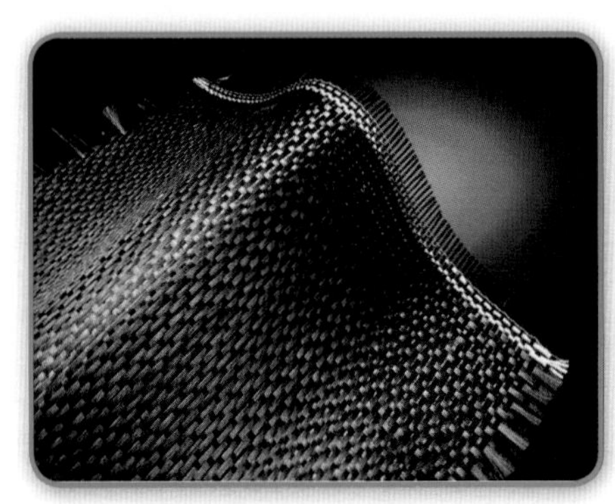

Figure 1.18: Carbon fibre weave

Medium density fibreboard

One of the most commonly used composite materials is medium density fibreboard (MDF). MDF is primarily made from wood waste (or specifically grown softwoods) in the form of wood chips, which are subjected to heat and pressure to soften the fibres and produce a fine, fluffy and lightweight pulp. This pulp is then mixed with a synthetic resin adhesive to bond the fibres and produce a uniform structure. It is then heat-pressed to form a fine-textured surface. After pressing, the MDF is cooled, trimmed and sanded. In certain applications boards are also laminated for extra strength. MDF can be worked like wood but with the added advantage that it has no grain to work with. It finishes well with a variety of surface treatments, and is available with a veneered surface for decorative effect.

Figure 1.19: An excellent strength-to-weight ratio is one reason why carbon fibre is used in high-performance racing cars

Support Activity

1. Many modern racing and mountain bikes have carbon-fibre frames. Why would carbon fibre be better for this use than traditional materials such as aluminium and steel?
2. Make a risk assessment of working with MDF in your workshop. Include the risks and control measures.

Composite	Uses	Advantages	Disadvantages
carbon fibre	sports equipment such as tennis raquets and fishing rods, bicycle frames and wheels, aircraft and vehicle components	• excellent strength-to-weight ratio • better tensile strength than steel alloys • fabric can be placed in different directions to provide strength in specific areas of structure • can be formed into complex and aerodynamic one-piece structures (distribute stress efficiently)	• very expensive material • only available in black (although surface finishes can be added to provide colour) • highly specialised manufacturing processes required • cannot be easily repaired as structure loses integrity • cannot be easily recycled
MDF	flat-pack furniture, general joinery work, moulds for forming processes	• less expensive than natural timbers • available in large sheet sizes and range of thicknesses • has no grain so no tendency to split • consistent strength in all directions	• requires appropriate finishes to seal surface fibres • swells and breaks when waterlogged (special treated variety required) • warps or expands if not sealed • contains urea-formaldehyde which may cause eye and lung irritation when cutting and sanding

Table 1.8: Uses, advantages and disadvantages of composite materials

Health and safety issues

MDF

Using any composite presents some potential hazards. Because of the fine fibres and synthetic resin adhesives used in MDF, great care must be taken when undertaking any form of cutting, drilling and especially sanding. You must use respiratory equipment and appropriate dust extraction, as the dust can irritate the skin, throat and nasal passages.

Carbon fibre

There are three areas of concern in the production and handling of carbon fibres: dust inhalation, skin irritation and the effect of fibres on electrical equipment. During processing, pieces of carbon fibre can break off and circulate in the air in the form of a fine dust. Although carbon fibres are too large to be a health hazard when inhaled, they can be an irritant, so you should wear a protective mask. To prevent skin irritation, use gloves or a barrier cream.

Stretch Activity

Use the internet to research the process of making a component out of carbon fibre. Make a step-by-step guide to producing that component.

ResultsPlus
Build Better Answers

Explain **two** reasons why a tennis player may prefer a racquet made from carbon fibre to one made from wood or aluminium. (4 marks)

■ **Basic answers (0-1 marks)**
Give one property of carbon fibre that makes it suitable for use in a tennis racquet.

● **Good answers (2-3 marks)**
State two properties of carbon fibre, with some reference to its advantage over either wood or aluminium.

▲ **Excellent answers (4 marks)**
Achieve full marks when they state two properties of carbon fibre and compare it with undesirable qualities in either wood or aluminium.

Modern and smart materials

Objectives

- **Describe** how modern and smart materials work.
- **Explain** the reasons for selecting modern and smart materials for different uses.
- **Understand** the structural composition of modern and smart materials.

What are modern and smart materials?

Modern materials have been developed through the invention of new or improved technologies, as a result of human intervention rather than naturally occurring changes.

Smart materials respond to differences in temperature or light (and other inputs such as pressure and sound) and change in some way as a result. They are classified as 'smart' because they sense the conditions in their environment and respond to those conditions. Smart materials appear to 'think', and some have a 'memory', as they revert back to their original state as soon as the input is removed.

Polymorph

Polymorph is one of a new generation of commercial plastics with unusual properties. At room temperature it is just as strong as a normal polymer, but at 60°C it becomes soft and mouldable. It can be reduced to a mouldable condition easily and quickly using hot water, and can be coloured or dyed using food-grade colourings.

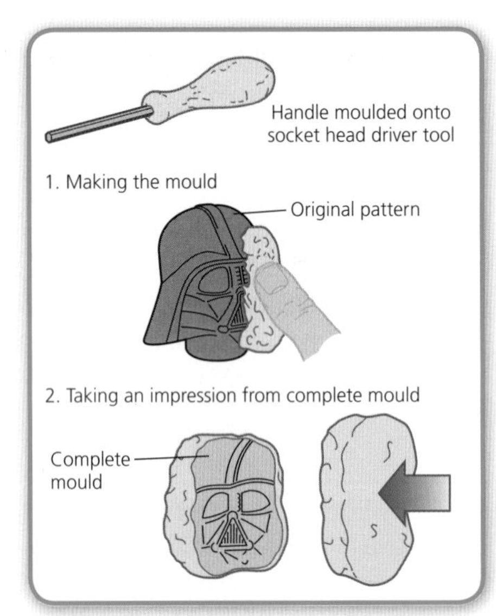

Figure 1.20: Polymorph can be used for one-off moulds or impression moulds

Once heated, polymorph becomes a soft, pliable material with the moulding properties of Plasticine. It has a range of uses for graphic products such as one-off mouldings for hand-held devices to test ergonomics, or vacuum-forming moulds and impression moulds that can be used to produce several copies of the same component. Once cooled, polymorph has the advantage of hardening like a plastic, whereas Plasticine remains soft. As a true thermoplastic, polymorph can be re-heated and thermoformed any number of times.

Thermochromic liquid crystals and film

Thermochromic liquid crystals have a number of uses including forehead thermometers, battery test panels and special printing effects for promotional items.

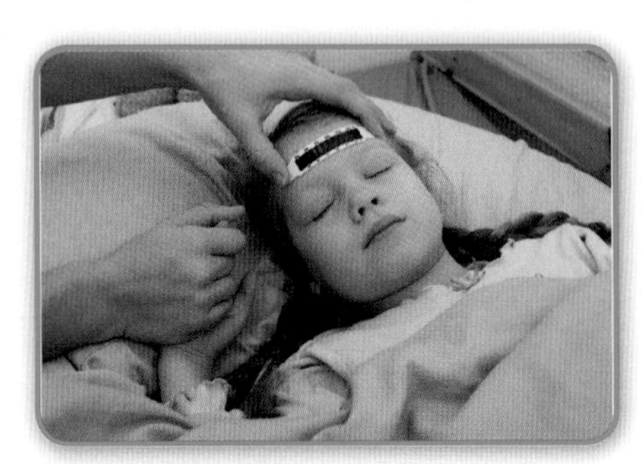

Figure 1.21: This forehead thermometer functions using thermochromic liquid crystals

For a forehead thermometer, a layer of conductive ink is screen-printed onto the reverse of the thermometer strip – the area that makes contact with the forehead. On top of the conductive ink is a layer of normal ink that conveys the temperature gauge colour bars. Finally, there is a thermochromic layer, which is black when cool. When you press the thermometer to your forehead, the temperature generated turns the thermochromic ink translucent. This reveals the temperature gauge colour bars that are printed in normal ink. Depending on your inner body temperature, most or all of the thermochromic ink will heat to the temperature needed to become translucent.

The same principle applies to battery test panels, where the electrical charge of the battery generates the heat required.

Other special printing inks are available to enhance printed materials especially for promotional use. Thermoreactive inks can be used to reveal graphics if a warm hand is placed over them or, conversely, if they are placed in a fridge – an ideal device for revealing a lucky winner on a promotional pack. As well as inks that react to changes in temperature, there are some inks that react to UV radiation in natural sunlight. These are known as photochromic inks.

Liquid crystal displays (LCDs)

Liquid crystals are organic, carbon-based compounds that can show both liquid and solid crystal characteristics. When a cell containing a liquid crystal has a voltage applied, and light falls on it, it appears to go 'dark'. This is caused by the molecular rearrangement within the liquid crystal.

In the case of a digital clock or wristwatch, an LCD display has a pattern of conducting electrodes that can display numbers using a seven-segment display. The numbers are made to appear on the LCD by applying a voltage to certain segments, which go dark in contrast to the silvered background.

As very small amounts of current are needed to power them, LCD displays are ideal for portable electronic devices such as mobile phones.

The rapid advance of LCD technology led to the full-colour LCD display commonly used in laptops. Here each pixel is divided up into three sub-pixels with red, green or blue filters. By controlling and varying the voltage applied, the intensity of each sub-pixel can create a range of over 256 colours. LCDs are now at the forefront of modern domestic appliance technology with even flatter, higher-resolution LCD televisions and computer screens.

Benefits of LCD displays

LCD displays:

- are thinner, so the TV, computer screen or laptop casing can be slimmer. This also means less material is used
- have a higher resolution, leading to better picture quality than traditional cathode ray tube (CRT) TVs, which tended to flicker due to their lower resolution
- use far less energy, saving both energy and costs
- weigh less than CRTs, so they are more portable.

Apply it!

You may be able to use polymorph in Unit 1: Creative Design and Make Activities.

Polymorph is a great modelling material for producing quick models during the development stage of the design activity.

Figure 1.22: Seven-segment LCD display

Figure 1.23: Full-colour LCD screens are ideal for laptop computers

Support Activity

Why are liquid crystal displays ideal for use on laptop computers?

Stretch Activity

Make a series of polymorph models to develop an ergonomic pen that is really comfortable to hold and use. You can mould the polymorph around an existing pen as a starting point.

Figure 1.24: Electronic paper is set to revolutionise display technologies

Electronic paper display (e-paper)

Electronic paper display (EPD), also known as electronic paper (e-paper), is a display technology designed to mimic the appearance of ordinary ink on paper. E-paper was developed to overcome some of the limitations of computer monitors: for example, the backlighting of monitors is hard on the human eye, but e-paper reflects light just like normal paper. E-paper is also easier to read at an angle than flat screen monitors. It is lightweight, durable and highly flexible compared to other display technologies, although it is not as flexible as paper.

E-paper is currently being developed for applications such as electronic books, capable of storing digital versions of many books. A major advantage of e-paper is that the pixels are extremely stable and require no power to maintain an image; as displays only draw on battery power when text is refreshed, they can display about 10,000 pages before the batteries need changing. Further technological developments will include electronic newspapers, where headlines can be constantly updated and animated images and video clips can be used.

How does e-paper work?

Each pixel point on the display is a tiny pit containing a small number of black and white beads, each one about as wide as a human hair. The white beads are positively charged and the black beads negatively charged. Each pit is topped with a transparent electrode and has two other electrodes at its base. Altering the charge on the base electrodes makes either white or black beads leap to the top of the pit, forming either a blank or black spot on the larger display. Making one base electrode positive and the other negative creates a grey spot. These black, white and grey spots can be used to create letters or pictures.

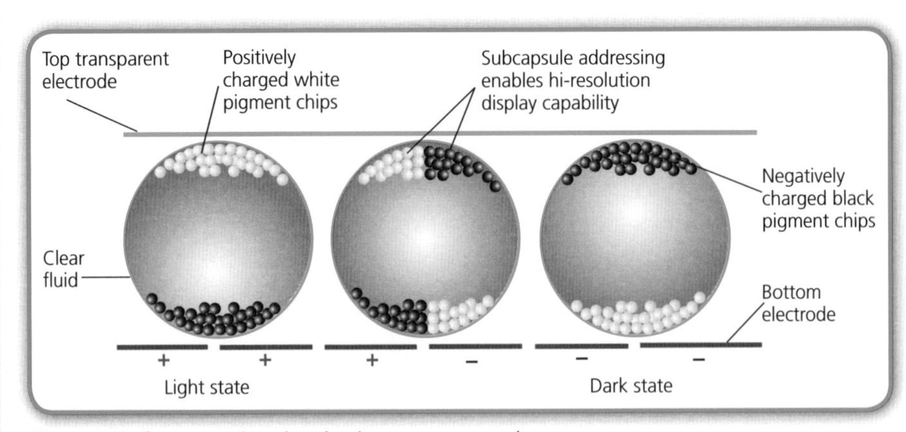

Figure 1.25: Cross-section showing how e-paper works

ResultsPlus
Build Better Answers

Explain **two** advantages of using an electronic paper display (EPD) rather than a liquid crystal display (LCD) in a new mobile phone. (4 marks)

Basic answers (0-1 marks)
Give only one feature of electronic paper display, with no justification of why it would be better than a liquid crystal display in a mobile phone.

Good answers (2-3 marks)
Give two relevant features of EPD that would make it suitable for a mobile phone, but some responses are not fully justified by comparing them with LCD.

Excellent answers (4 marks)
Achieve full marks when both beneficial features of EPD are fully justified compared to LCD and applied to use in a new mobile phone.

Transdermal prescription drug patches

A transdermal patch is a medicated adhesive patch that delivers a specific dose of medication through the skin and into the bloodstream. The patch contains a drug reservoir sandwiched between a non-permeable back layer, and a permeable adhesive layer that attaches to the skin. The drug leaches slowly out of the reservoir, releasing small amounts of the drug at a constant rate for up to 24 hours.

One advantage of these patches is that they provide a controlled release of the prescription drug into the patient, unlike the usual ways of taking drugs by mouth or injection. A major disadvantage is that the skin is a very effective barrier, so some prescription drugs, such as insulin for diabetes, cannot pass through the skin. Some prescription drugs can only be used in patches if they are combined with substances such as alcohol, to increase their ability to penetrate the skin. However, many prescription drugs can be delivered by transdermal patches, and they can be used to help people quit smoking, to relieve chronic pain and even as slimming aids.

Component	Use
foil backing	non-permeable layer protects the patch from the outer environment
prescription drug reservoir	holds a precise amount of the prescription drug, in direct contact with the membrane
membrane	permeable layer that controls the release of the prescription drug from the reservoir
adhesive	bonds the components of the patch together as well as sticking the patch to the surface of the skin

Table 1.9: The main components of a transdermal patch

Benefits of transdermal patches as an alternative to injections

A transdermal patch:

- provides a controlled release of the medication into the patient over a period
- delivers the medication painlessly, as it is simply absorbed through the skin
- is a 'clean' technology, where there are no dangerous needles to dispose of.

Support Activity

1 Why would transdermal patches be good for giving children their medicine?

2 Why would electronic paper be good for your exercise books at school or college?

Explain your answers.

Stretch Activity

Use the internet to research the range of products currently available or under development that make use of electronic paper displays.

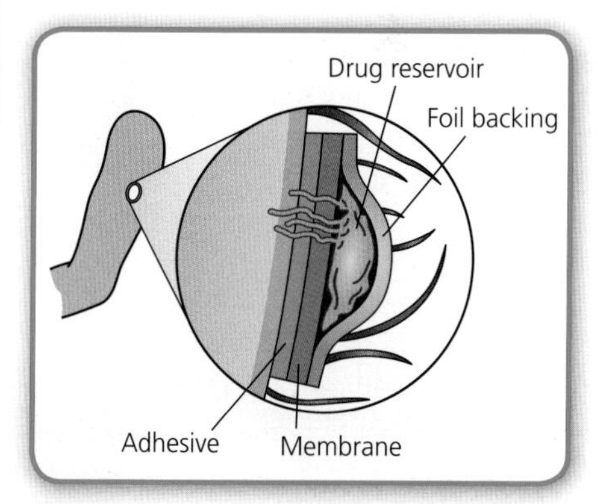

Figure 1.26: Patches control the release of prescription drugs into the bloodstream

Components

Objectives

- **Describe** the use of a range of binding methods and technical drawing equipment when producing graphic products.
- **Explain** the reasons for using different binding methods for specific graphic products.

During this course you will undoubtedly be asked to complete a working drawing of a product you have designed. To achieve this you will have to use a range of technical drawing equipment to ensure that you effectively communicate your idea. You may also produce printed graphic products and documents that need to be bound together.

Binding methods

Binding is a process used to fasten or hold together a number of printed sheets. Many products are bound, such as magazines, books, reports and brochures. Binding can range from the simplest forms, such as stapling or ring binding, to fully automated processes. There are various methods of binding to choose from, the choice depending on the specific application. Factors such as aesthetic considerations, the quantity of paper to be bound and the cost will determine which process is used.

Method	Diagram	Uses	Advantages	Disadvantages
saddle-wire stitching the simplest method of binding by stapling the pages through the fold		brochures, weekly magazines, comics	• ideal for signature-feed processes (folded pages) • printed materials can be laid flat to read • relatively inexpensive when commercially produced	• lower-quality visual appearance • not durable as centre pages can easily fall apart
spiral and comb binding pages are punched through with a series of holes along the spine; a spiralling steel or plastic band is inserted through the holes to hold the sheets together		business reports/ documents	• relatively inexpensive when produced commercially • ideal for binding multiples of single sheets of paper without folds • fairly good quality visual appearance • printed materials can be laid flat to read	• not durable as document can easily fall apart or tear down perforations
perfect binding pages are held together and fixed to the cover by means of a flexible adhesive		paperback books, glossy monthly magazines, catalogues	• better presentation and visual appeal with printable spine rather than staples • better quality – puts all the pages or signatures together, roughens and flattens the edge, then a flexible adhesive attaches the paper cover to the spine • glued spine provides longevity for a monthly magazine	• expensive commercial process
hard-bound or case-bound usually combines sewing and gluing to create the most durable method of commercial binding		hardback books, quality presentations such as school yearbooks	• stiff board used on the cover to protect the pages • high quality, professional binding method • extremely durable	• very expensive commercial process

Table 1.10: Binding methods for paper and board

Technical drawing equipment

When designing your graphic products, you will have to use a range of technical drawing equipment. This will range from pencils for sketching initial design ideas to a drawing board, compass and set square for constructing a final working drawing.

Support Activity

Produce a third angle orthographic drawing of a familiar product using a drawing board and technical drawing equipment.

Name	Equipment	Description
pencil		for drawing and sketching • hard pencils (H to 9H) are used for technical drawing as they can be sharpened and retain a fine point for accurate lines • soft pencils (9B to HB) are used for sketching and shading as the lead is richer and darker
set square		a drafting aid available in 45° or 30/60°, used with a drawing board to produce technical drawings • 45° set squares can be used for oblique drawings and cross-hatching of sectional drawings • 30/60° set squares can be used for isometric drawings (30°) and planometric/axonometric drawings (60°)
compass		for drawing circles and arcs and marking measurements • spring-bow compasses are ideal for drawing small-diameter circles • pencil compasses are ideal for larger-diameter circles
ruler		a straight edge for drawing lines, with a scale (mm) for measuring
circle/ellipse template		• a circle template offers a quick and effective way of drawing smaller diameter circles without the need for a compass • an ellipse template enable circles to be drawn in isometric view without the need for complex technical constructions
flexicurve/ french curves		• french curves provide a means of repeating a particular curve without having to technically construct it again • a flexicurve is a plastic strip with a lead core that can be bent into any desired curve and will retain that shape
drawing board		a flat surface to attach paper securely to with a parallel motion (sliding rule) that aids technical drawing

Table 1.11: Technical drawing equipment for designing graphic products

Stretch Activity

Produce an eight-page A5 booklet (2 × A4 sheets folded in half) about a hobby or interest that appeals to you. Use a computer to design it and practise your desktop publishing (DTP) skills. Use a long-armed stapler to staple through the spine, to simulate saddle-wire stitching.

Know Zone
Chapter 1 Materials and Components

As a designer, you need to know about the properties of a wide range of materials and components so that you can make informed choices about their use in certain products.

You should know...

1 about the following about materials:

Materials	Properties	Advantages and disadvantages	Uses/applications	Structural composition
Paper and board: Cartridge paper, Tracing paper, Folding boxboard, Corrugated board, Solid white board, Foil-lined board	✓	✓	✓	×
Metals: Steel, Aluminium, Tin	✓	✓	✓	×
Polymers: Acrylic, PET, PVC, Polypropylene, Rigid polystyrene, Expanded polystyrene, Styrofoam	✓	✓	✓	×
Glass	✓	✓	✓	×
Woods: Jelutong, Balsa, Pine	✓	✓	✓	×
Composites: Carbon fibre, MDF	✓	✓	✓	×
Modern and smart materials: Polymorph, Thermochromic liquid crystals, Electronic paper, Transdermal patches	✓	✓	✓	✓

2 about the following about components:

Components	Processes	Advantages and disadvantages	Uses
Spiral/comb binding	✓	✓	✓
Saddle-wire stitching	✓	✓	✓
Perfect binding	✓	✓	✓
Hard/case binding	✓	✓	✓
Pencils	×	×	✓
Set squares	×	×	✓
Compasses	×	×	✓
Rulers	×	×	✓
Circle/ellipse templates	×	×	✓
French curves/flexi-curves	×	×	✓
Drawing boards	×	×	✓

Key terms

Many questions relating to materials and components will require you to know their properties and apply them for use in different products.

Property	Meaning
Aesthetic	The look and feel of materials such as colour, style and texture. For example, solid-white board is used for luxury packaging because it has a high-quality visual appearance.
Functional	Qualities that materials must have in order to be 'fit for purpose' such as strength, weight and durability. For example, carbon fibre is used for many products where a high strength-to-weight ratio is essential.
Mechanical	A material's reaction to physical forces such as plasticity, ductility, hardness and malleability. For example, aluminium is soft and highly malleable and can therefore be formed into tubes for drinks cans.

ResultsPlus
Maximise your marks

Grade E-C range question: Explain **one** property of Styrofoam™ that makes it suitable for making models. (2 marks)

Student answer	Examiner comments	Build a better answer
Styrofoam is soft. (1 mark)	'Softness' is an appropriate property of Styrofoam™ but the learner does not go on to explain why it is suitable for making models.	Styrofoam™ is a soft material (1 mark) so it can be easily cut and shaped for model making. (1 mark)

Overall comment: Many learners do not achieve full marks in 'explain' questions. Although they make a valid point they do not then go on to justify that point. It is always worth having an 'educated guess' because an examiner may be able to give you the benefit of the doubt even if your response is not exactly the same as the one stated in the mark scheme.

Grade C-A range question: Explain **two** advantages of using PET bottles, rather than glass bottles, for a consumer. (4 marks)

Student answer	Examiner comments	Build a better answer
1 *PET does not smash like glass if you drop it.* (1 mark)	The learner's first response is fine but does not go on to justify it. Why is 'not smashing like glass' an advantage?	PET will not shatter like glass if dropped (1 mark) therefore it is safer to use. (1 mark)
2 *PET is lighter than glass* (1 mark) *so it is easier to carry home from the shop.* (1 mark)	This is a perfectly acceptable response because it makes a point and then justifies why.	PET is a lightweight material for packaging (1 mark) therefore it is easier for the consumer to carry than a glass bottle. (1 mark)

Overall comment: Don't forget that 'explain' questions carry 2+ marks so always need a valid point plus justification.

Chapter 2 Industrial and commercial processes

Scale of production

Objectives

- **Describe** the main features of the three scales of production: one-off, batch and mass.
- **Explain** the advantages and disadvantages of using each scale of production.
- **Understand** the social issues relating to each scale of production.

The term 'scale of production' refers to the quantity or numbers of a product made, depending on the needs of the customers. A manufacturer will decide to use either one-off, batch or mass production processes based on:

- the volume or quantities of products required
- the types of materials used to make the products
- the type of product being manufactured.

One-off production

One-off production is often referred to as job production and includes 'tailor-made' and customised designs. One-off production creates a single product at a time, manufactured to a client's specification, often at a high cost. This is because a premium has to be paid for:

- any unique features
- more expensive or exclusive materials
- time-consuming, hand-crafted production and finishing.

Craftsperson

The modern craftsperson is highly skilled, but will often work on jobs designed by other people. For example, a company needing a vacuum-formed tray for a cosmetics gift box would require a mock-up of the final product before it could consider the box for full production. A pattern maker may be commissioned to make a wooden mould of the tray and vacuum-form it, to test its fitness for purpose.

The role of the pattern maker is increasingly being taken over by CAD/CAM and rapid prototyping (which you will look at in more detail on pages 42–43). With this technology you can use a relatively simple computer software to construct a 3D computer model of a mould. The data for this mould can then be downloaded into a computer-controlled machine, which can cut the mould to shape. The time taken to produce the mould using these techniques can be drastically reduced.

With this in mind, manufacturers will often use rapid prototyping to aid in the development of a product, rather than employ an expensive craftsperson.

Model makers

Specialist model makers are commissioned to make architectural models or interior layouts for proposed buildings. A detailed 3D model can communicate much more about a building than any 2D drawing ever could.

Model making is a highly skilled job requiring great attention to detail, and is not easily done by computers. Because of this, both design and production costs are high.

Figure 2.1: A one-off architectural model

Batch production

Batch production involves the manufacture of identical products in specified quantities, which can vary from tens to thousands. Key features of batch production include:

- having flexible tooling, machinery and workforce – this enables fast turnaround, and means that production can quickly be adapted to manufacture a product to meet a new demand
- making use of flexible manufacturing systems (FMS) so that companies can be competitive and efficient
- the use of computer-integrated manufacturing (CIM) systems involving automated machinery so that production down time (time when not manufacturing) can be kept to a minimum.

Batch production results in a lower unit cost than one-off production. Economies of scale in buying materials can save costs, so that identical batches of consistently high-quality products can be manufactured at a competitive price.

Flexible manufacturing systems (FMS)

Flexible manufacturing systems (FMS) are intended for batch production work where the time taken to set up new tools and machines for different jobs is crucial. The aim is to make the most efficient use of manufacturing processes while remaining flexible enough to respond quickly to the needs of other jobs. The 'flexible' factory uses more flexible equipment, involving ICT systems that have the ability to perform more than one task on a wide range of products.

Controlling workflow

The term 'workflow' describes the tasks or stages required to produce a final product. In a large commercial printing company, it is essential that a range of printed materials can be produced for different customers at the same time.

Figure 2.2 shows the workflow in a printing company using two 8-colour lithographic printing presses and two 4-colour presses. At any one time, the different presses are being used for different print jobs: for example, the 8-colour presses may be printing a batch of 5000 full-colour brochures for one customer while the 4-colour presses are involved with the mass production of a leaflet for another customer.

The print floor is organised in a way that allows paper stock to be transformed into the final printed product as efficiently as possible. Large stock deliveries, transported on lorries, enter the factory via a loading bay. Pallets of paper stock are then unloaded and stored using fork-lift trucks. Printing plates are designed, made up and put onto the presses. The paper is taken out of stock and fed into the press for full-colour printing. The printed paper is then transferred to the finishing area where it is folded, stitched and trimmed. When ready, the final printed products are counted and put on pallets ready for delivery or transfer to a warehouse for storage.

Apply it!

You will be required to produce a one-off model in your Unit 1 Make Activity.

Your final model should be of a high quality, so you will have to use a range of skills and processes with precision and accuracy, and make sure that you finish your model well.

Support Activity

1 Why is an Aston Martin sports car more expensive than a Ford Fiesta?

2 Why must a modern business be as flexible as possible?

Stretch Activity

Discuss the future of craftspeople in a world driven by technology.

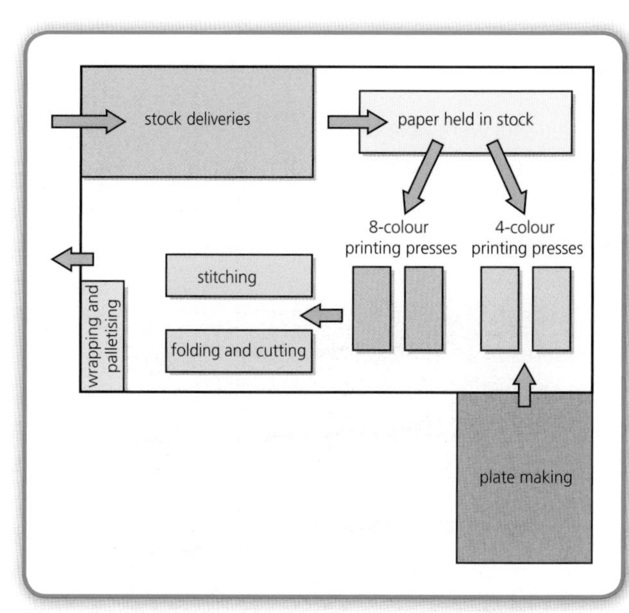

Figure 2.2: Workflow on the print floor during batch production

Figure 2.3: Fizzy drinks are mass-produced using production lines

Mass production

Mass production of most consumer products makes use of efficient, automated manufacturing processes and a largely unskilled workforce. Production is divided up into specific tasks, with labour to match.

Mass-produced products are designed to follow mass-market trends, so the product appeals to a wide national and international target market. Production planning and quality control (QC) in production enable the manufacture of identical products. Production costs are kept as low as possible by using production lines or manufacturing cells (see below), so that the product will provide value for money.

In-line production and assembly

In-line production and assembly is the traditional approach to mass production. It utilises low-cost, unskilled labour supported by semi-skilled, flexible people who can change tasks as required to make sure the assembly line runs smoothly.

Manufacturing cells

Manufacturing cells are a way of organising workers into semi-autonomous and multi-skilled teams called work cells, which manufacture complete products or complex components. A manufacturing cell is much more flexible and responsive to changes in the quantities of products needed than the traditional mass-production line; a manufacturing cell can also manage its own production schedules, quality control, equipment maintenance and other manufacturing issues more efficiently.

It is extremely important that members of the cell work together as a team and share responsibility for the cell output. Manufacturing cells vary from small teams of people to fully automated cells with computer-controlled machines and robots.

Just-in-time (JIT) stock control

Traditionally, products have been manufactured according to an agreed schedule or plan governed by the availability of materials and components. Just-in-time (JIT) stock control is based on production being 'pulled' along in response to customers' orders and the requirements of the manufacturing process.

The aim is to rid the system of stock at every opportunity (stock includes materials and components as well as finished products held in storage). Keeping stock in storage wastes money and takes up valuable space in the factory. JIT is totally dependent on the supply of materials and components arriving just in time.

Productivity

Modern mass-production systems ensure a company's high productivity – that is, the speed and efficiency with which a company turns raw materials into a final product. The most common measure of productivity is the amount of products a single worker can produce. The more products that a single worker can produce, the lower the labour costs; the lower the labour costs, the higher the potential profit.

ResultsPlus
Exam Question Report

Describe how an automatic production system enables faster production. (3 marks)

How students answered

Many students only stated one feature of an automatic production system but failed to provide reasons why this feature enables faster production.

30% 0-1 marks

Most students were able to state two features of automatic production that enable faster production, or fully justified one feature.

45% 2 marks

Some students could identify at least two features of an automatic production system and justify why they enabled faster production. Alternatively, some students achieved full marks by identifying one feature, giving a detailed justification with two clear marking points.

25% 3 marks

Choosing the scale of production

Scale of production is an important factor to be considered when developing any graphic product. It has an impact on all design and manufacturing decisions, including:

- the number of products or units manufactured
- the choice of materials and components
- the manufacturing processes, speed of production and availability of machinery and labour
- production planning, the use of JIT and stock control, including the use of ICT systems
- production costs, including the benefits of bulk buying, the use of standard components and eventual retail price.

Figure 2.4: Injection moulding using a fully automated manufacturing cell

Stretch Activity

Watch a DVD or TV programme that shows a product being made from start to finish. Note down the key stages in its manufacture.

Scale of production	Uses	Advantages	Disadvantages
one-off	prototype and architectural models, shop signage, vinyl stickers for commercial vehicles, etc.	• made to exact personal specifications • high-quality materials used • highly skilled craftsperson ensures high-quality product	• usually an expensive final product (although a one-off poster from a colour printer may be cheap) • usually labour-intensive and slow process (although e.g. digital printing of photographs takes seconds)
batch	commercially printed materials, such as magazines and newspapers	• flexibility in adapting production to another product • fast response to market trends • identical batches of products produced • efficient manufacturing systems can be employed • good economies of scale in bulk-buying of materials, depending on batch size • lower unit costs	• poor production planning can result in large quantities of products having to be stored, incurring storage costs • frequent changes in production can cause costly re-tooling, which is reflected in the retail price
mass	electronic products such as mobile phones and games consoles, graphic products such as clothes labels, tickets, packaging	• highly automated and efficient manufacturing processes • specialisation of workforce to specific tasks • rigorous quality control ensures identical goods • excellent economies of scale in bulk-buying of materials • increased production leads to quick recovery of set-up costs • low unit costs • low labour costs	• low skilled workforce – low wages, repetitive nature of tasks leading to job dissatisfaction (note that, from a printing perspective, the skills involved in batch and mass are the same) • ethical concerns of manufacturing in developing countries, e.g. sweatshops • high initial set-up costs due to expensive machinery and tooling • inflexible – cannot respond quickly to market trends

Table 2.1: Uses, advantages and disadvantages of scales of production

Modelling and prototyping

Objectives

- **Describe** the processes of block modelling of MDF and Styrofoam™ and rapid prototyping using stereolithography (SLA) and 3D printing (3DP).
- **Name and describe** the use of common workshop tools, equipment and components used in making graphic products.
- **Explain** the reasons for using rapid prototyping in the development of graphic products.

Block modelling

Block modelling helps the designer to determine shape, dimensions and surface details by making 3D models of the proposed product. Block models can be extremely useful in determining the ergonomic factors of many products. By making a number of block models of varying shapes and sizes it is possible for designers to literally get a 'feel' for the product. It will soon become apparent which designs are aesthetically pleasing or 'user friendly' and are worth developing – something that 2D images struggle to achieve.

Block modelling of MDF

The most important role that medium density fibreboard (MDF) has to play in a graphic product is in the manufacture of three-dimensional (3D) models and moulds for vacuum forming. MDF is an ideal material for producing a high-quality model because it:

- can be cut and shaped easily with a range of hand tools, as it has no grain
- has an excellent surface finish when sanded smooth
- can be spray-painted, once sealed, to achieve a high-quality, professional finish.

When designing a product such as a mobile phone or games controller, MDF can be formed into the smooth, streamlined shapes that are essential for modern-looking products.

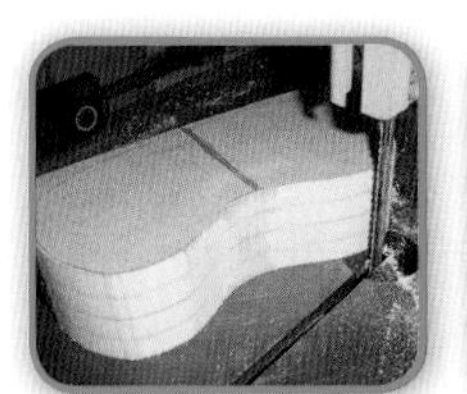

Figure 2.5: Cutting MDF block to a rough shape using a bandsaw

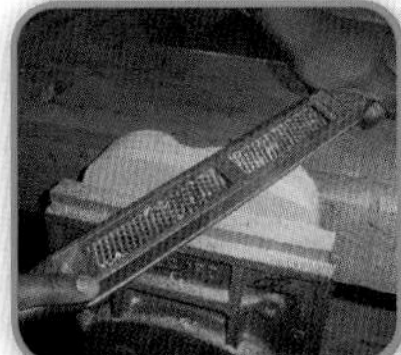

Figure 2.6: Shaping MDF using a surform

To create an MDF model, mark the plan and side profiles of the rough shape on a block, which your teacher can cut on a bandsaw. Shape the cut MDF using tools such as surforms, rasps and files to achieve the desired product styling, then sand the model smooth using various grades of glasspaper. To achieve a quality finish, use an acrylic spray primer, sanding it back gently using wet and dry paper. Good-quality acrylic car paints are available in a wide range of colours, and you can use these to apply a professional-looking top coat.

MDF can also be used for the production of moulds for vacuum forming: for example, to produce the transparent 'blister' in blister packaging.

Figure 2.7: Achieving a professional-looking finish using acrylic car paints

Figure 2.8: Finished MDF model

A mould can be cut and shaped to create interesting shapes in much the same way as producing a product model. MDF is the most suitable wood because it has no grain, so the mould will not leave an imprint on the vacuum-formed plastic shape. It is important that the completed mould is very smooth, has slightly angled sides (usually 5°) and rounded or 'radiused' corners and edges. This will ensure that the mould can be easily removed once vacuum formed.

Block modelling of Styrofoam™

As Styrofoam™ can be easily cut and shaped with a range of hand tools, it is ideal for block modelling. 3D models can be produced relatively quickly, especially at the development stage where models can be used to test and refine design ideas. Final models can be sanded extremely smooth and can be painted to give a good quality finish if desired. However, a more professional finish is accomplished by coating the Styrofoam™ with several layers of plaster, sanding down between coats and then spraying with acrylic car paints.

Workshop tools, equipment and components

The tables below show some of the tools that you could use when making your models and explain their uses.

Cutting	Name and use
	bandsaw heavy-duty electric saw for cutting most sheet materials
	vibrosaw bench-mounted electric saw for cutting thin sheet materials
	hot wire cutter a taut, heated metal wire that cuts through Styrofoam™ and expanded polystyrene (EPS)
	coping saw saw for cutting curves out of sheet materials including thin wood and acrylic
	pillar drill bench-mounted electric drill for drilling holes in most materials

Table 2.2: Cutting tools for use in block modelling

Shaping	Name and use
	surform roughly shaping soft materials, e.g. laminated MDF block models
	rasp shaping soft materials
	file creating a smooth finish on acrylics and MDF

Finishing	Name and use
glasspaper	making a smooth surface finish on woods and Styrofoam™
wet and dry paper	making a smooth surface finish on acrylic and MDF
sanding sealer	sealing the porous surface of MDF prior to painting
spray paint	applying a professional-looking finish to MDF once sealed

Tables 2.3 and 2.4: Shaping and finishing tools for use in block modelling

ResultsPlus
Build Better Answers

Using notes and sketches, explain the process of producing a prototype using stereolithography. (4 marks)

■ **Basic answers (1-2 marks)**
Include a basic sketch with limited annotation showing a machine with a laser and a model sliced into several layers.

● **Good answers (3 marks)**
Include a more detailed sketch, with some good notes explaining the process.

▲ **Excellent answers (4 marks)**
Include a couple of sketches with descriptive, detailed notes, clearly showing the model being built up in layers.

Rapid prototyping

The need for manufacturing industries to cut down on the time and costs involved in developing a new product has led to the development of rapid prototyping.

When you make a block model you will usually remove material when you shape it (e.g. wood dust) – and producing this waste material is not efficient. Rapid prototyping (RPT) is a much more efficient, computer-controlled process that builds up the model from scratch instead of taking away materials from a block.

The advantages of using RPT in the development of products include:

- fast entry to market due to a reduction in lead time – the time between the initial design ideas and the actual product being sold
- reduced development time, which saves money
- produces complex, intricate shapes accurately directly from CAD data
- accurate testing of models as materials are more representative of the final product: for example, this process produces an actual polystyrene casing rather than a MDF or Styrofoam™ block model of one.

Two RPT processes commonly used by manufacturing industries are stereolithography (SLA) and 3D printing (3DP).

Rapid prototyping using stereolithography

Stereolithography involves the creation of 3D objects using laser technology to solidify liquid polymers or resins. Specialist software applications can be downloaded onto a stereolithography machine, so that 2D CAD drawings can be converted into 3D models.

The process is based on the computer 'slicing' the virtual 3D object into hundreds of very thin layers (typically 0.125-0.75 mm thick) and transferring the data from each layer to the laser. The laser draws the first layer of the shape onto the surface of the resin, causing it to solidify. This layer is supported on a platform that moves down, enabling the next layer to be drawn. This process of drawing, solidifying and moving down quickly builds the shape up, one layer on top of another, until the final 3D model is achieved.

Stereolithography prototypes are often delivered within three to five days of receiving the client's data, so this process saves both time and development costs. So far only a few companies have this technology.

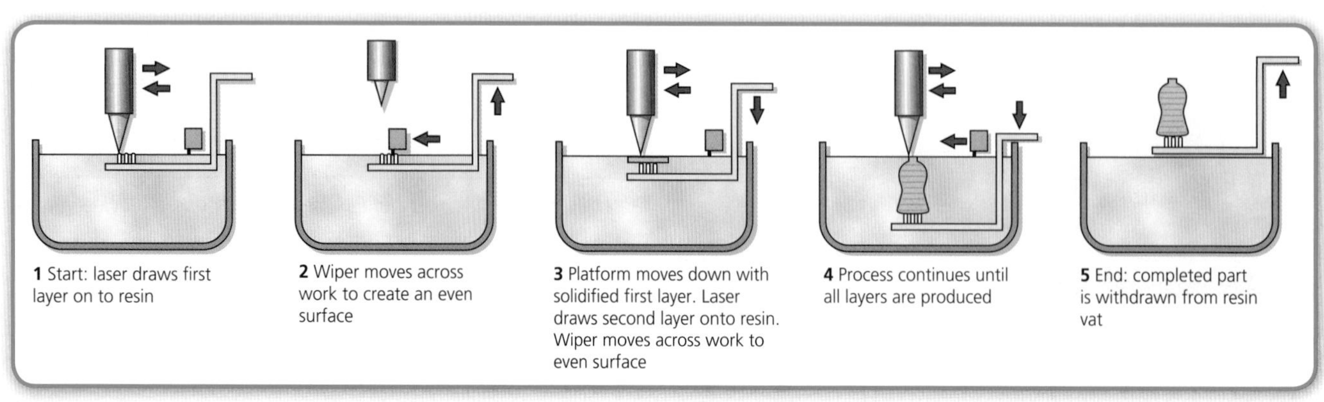

Figure 2.9: The process of stereolithography

3D printing

3D printing gives designers and product development teams rapid prototyping technology in the office, so that they can produce 3D models in minutes. 3D printers have become financially accessible to small and medium-sized businesses, including some schools, taking prototyping out of heavy industry and into the office environment.

CAD software sends the 3D image to the 3D printer and the item is 'printed' layer by layer in a range of materials. Concept models can be produced quickly and with working parts using a desktop-size 3D printer. 3D printers offer speed, low cost and ease of use, which makes them suitable for visualising designs during the development stage of the design process.

Like stereolithography, 3D printing works by converting a 3D CAD file into thin cross-sections. This information is then sent to the 3D printer, which starts to deposit material layer by layer, creating a 3D object. One variation of 3D printing involves an inkjet printing system. Layers of a fine powder are bonded by 'printing' an adhesive from the inkjet printhead in the shape of each layer, as sliced up by the CAD file. This system allows the printing of full-colour prototypes and is also a fast method of producing a complex prototype. Another similar system feeds a liquid photopolymer through an inkjet-type printhead to form each layer of the model. These photopolymer phase machines use an ultraviolet (UV) floodlamp mounted in the printhead to solidify each layer as it is deposited.

Some 3D printers can create models out of a tough polymer called ABS including working parts that enable a designer to test form, fit and function. This higher level of precision enables parts to actually click-fit together. ABS produces durable and functional prototypes that can withstand rigorous testing and will not warp, shrink or absorb moisture.

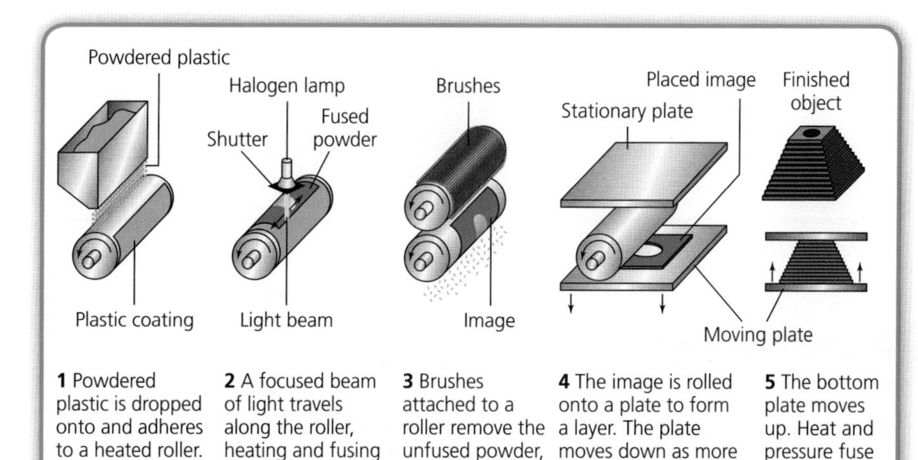

Figure 2.10: 3D printing process

Apply it!

If your school or college has a 3D printer, then why not use it in your Unit 1: Design and Make Activities?

- Design Activity: use it at the development stage to test and refine your ideas.
- Make Activity: use it to produce some components on your overall model. Note that CAM can only be used up to 50 per cent.

Support Activity

1 Why would a designer use the services of a rapid prototyping company when developing a product?
2 Why is there still a need for craftspeople such as model makers who make models primarily by hand?

Stretch Activity

Use the internet to research the different types of rapid prototyping services provided by a range of companies. Look at the types of products that have been produced.

Forming techniques

Objectives

- **Describe** the processes of blow moulding, injection moulding, vacuum forming and line bending thermoplastics.
- **Describe** quality control (QC) inspection and testing procedures when thermoforming products.
- **Describe** the process of printing directly onto thermoformed products.
- **Explain** the advantages and disadvantages of using these methods for batch and mass production of graphic products.

Thermoforming products

Many products are batch- and mass-produced using thermoforming techniques to mould and shape polymers. Thermoplastics are commonly used because they can be easily moulded and any waste produced can be recycled and used again in the process.

Blow moulding and injection moulding can produce large quantities of identical products very quickly. Vacuum forming is also widely used in industry and is ideal for producing batches of similar products within schools. In the school workshop, line bending of acrylic enables you to produce high-quality products or models with some accuracy.

Blow moulding

In the blow-moulding process, a hollow thermoplastic tube or parison is extruded or forced out between a split two-piece mould and clamped at both ends. Hot air is blown into the parison, which expands to the shape of the mould, including relief details such as threads and surface decoration. Once the polymer solidifies, you eject the product by opening the split mould. Blow-moulded containers do not have to be symmetrical and can incorporate handles, screw threads and undercut features.

Figure 2.11: Fizzy drink bottles are blow-moulded

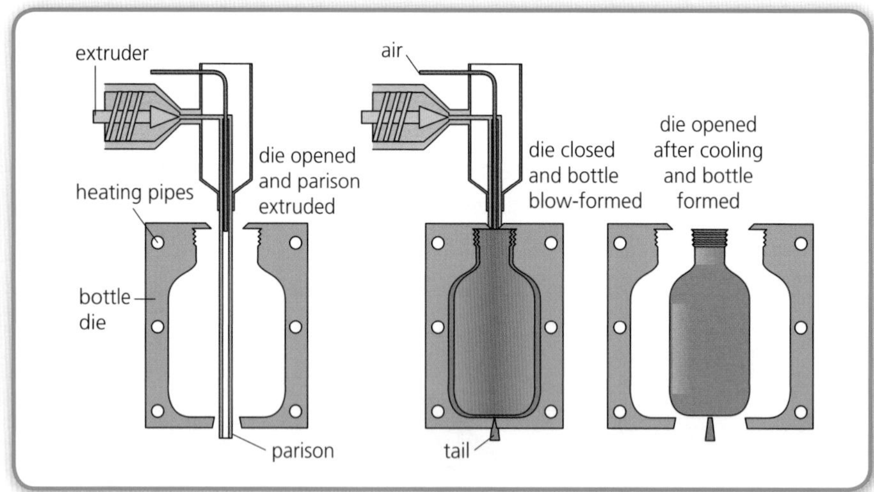

Figure 2.12: The blow-moulding process

Dome blowing

Dome blowing is the process of forming domes, spheres and oval shapes, usually out of acrylic. The sheet of acrylic is softened in an oven and transferred to a dome-blowing machine, where it is clamped under a circular ring. Air pressure is applied, which blows the material upwards and forms it into a perfect dome shape. Commercial dome blowing can produce perfect domes of a maximum diameter of two metres without distortion. This process is used to produce signage and point-of-sale displays.

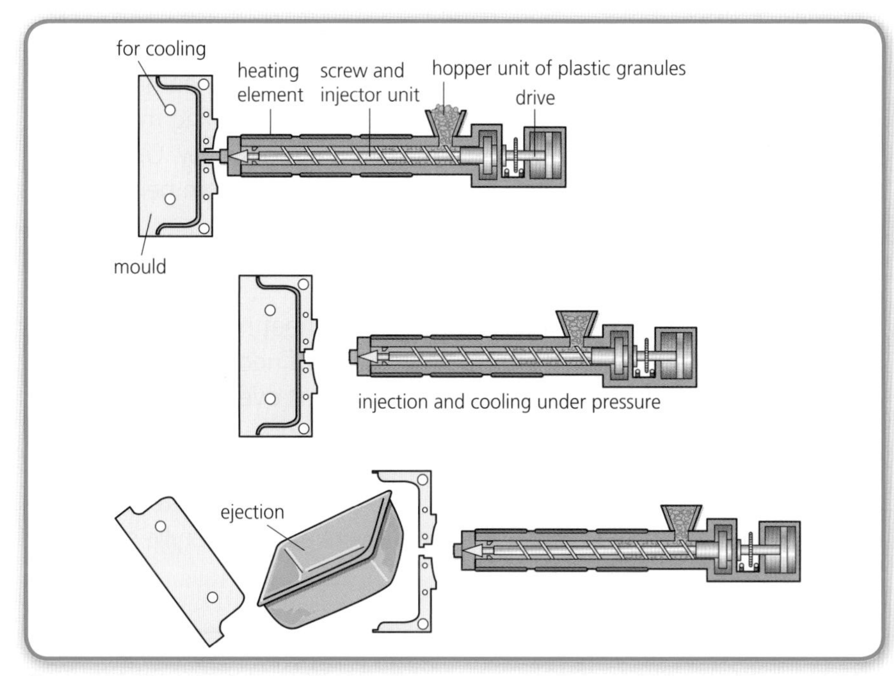

Figure 2.13: The injection-moulding process

Injection moulding

In the injection-moulding process, an expensive mould is injected with a liquid polymer, made by heating thermoplastic granules. Once the polymer cools and solidifies, the formed product is ejected. Injection moulding is suitable for complex shapes with holes, screw fittings and integral hinges.

Quality control (QC) inspection and testing

As many of these thermoforming processes are used for high batch or mass production, it is important that each product is identical in quality. Imperfect products have to be scrapped, and while many can be recycled, wastage still costs a company time and money. Companies will use several different types of machine to inspect batches of products at different stages of their production including laser measurement and ultrasonic testing.

- **Laser measurement** uses a series of lasers to measure the outside dimensions of the product as they move down a conveyor belt. This is a fast, accurate and repeatable process as lasers are computer-controlled and data is available immediately.
- **Ultrasonic testing** is used to check the wall thickness of a hollow product by sending high-frequency sound waves at the product as it moves down a conveyor belt. The waves will bounce back differently as they hit different materials and different thicknesses. The computer analyses the time it takes the sound waves to bounce back in order to calculate the wall thickness of the product accurately.

Support Activity

Make a list of products that are made using blow moulding and injection moulding. Collect images of these products for your revision notes so that you can give examples in an exam.

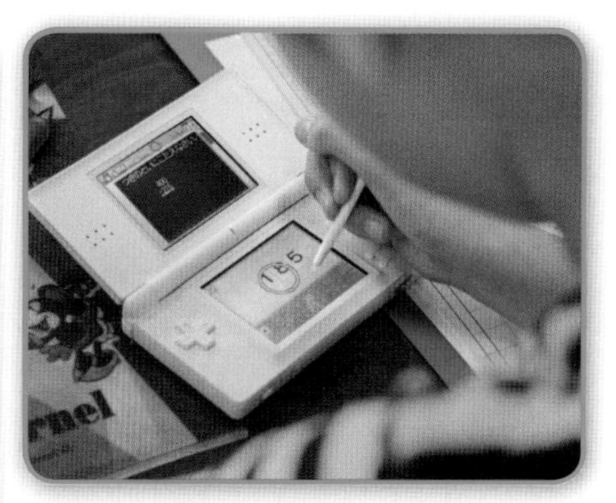

Figure 2.14 The casings for electrical products are usually injection-moulded

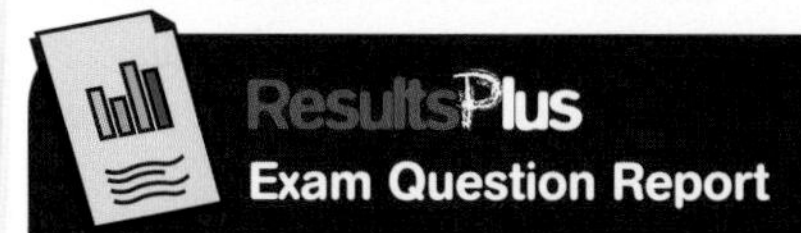

Explain two reasons why the injection-moulding process is suitable for mass production. (4 marks)

How students answered

Many students only stated one appropriate reason with no justification. Many of these students made simple statements such as 'cheaper', 'faster' and 'easier', which gained no marks at all.

36% 0-1marks

Most students could include two valid reasons why injection moulding is a suitable mass-production process. However, they only justified one reason in sufficient detail.

51% 2-3 marks

Some students correctly identified two appropriate reasons, fully justifying each. Remember: in an 'explain' question, you have to make a valid point and then go on to justify it. You cannot achieve full marks without any justification.

13% 4 marks

Stretch Activity

Search the internet for animated diagrams of thermoforming processes so that you understand the process in detail.

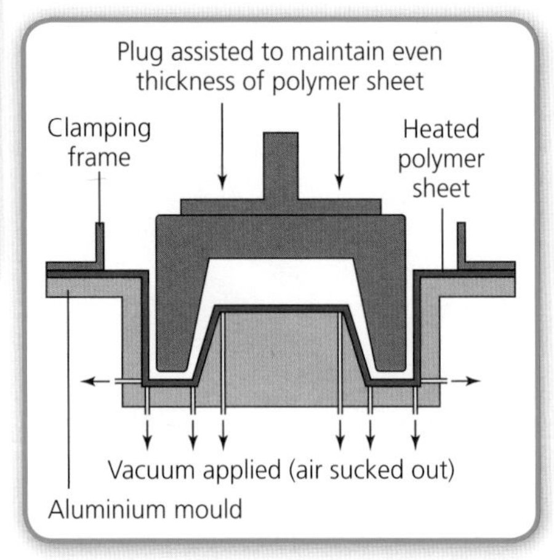

Figure 2.15: Commercial vacuum forming uses a plug to help form an accurate shape

Vacuum forming

In vacuum forming, a thermoplastic sheet is clamped and then heated, blown and stretched. Air is sucked out of the vacuum-forming machine to force the softened sheet over a mould pushed up from below. Once the polymer has cooled, it solidifies and cold air is blown up from below to release the formed product.

Line bending

Line bending involves heating a thermoplastic sheet material such as acrylic over a strip heater, along a narrowly defined line, until it becomes soft and pliable. It is then usually bent over a former. Thermoplastics may be bent to any angle by using a jig or a former or, if the angle is not critical, simply by bending the thermoplastic sheet by hand and then holding it until it cools. This process allows items such as display stands, leaflet dispensers, detail pockets and acrylic signs to be made from flat-sheet material.

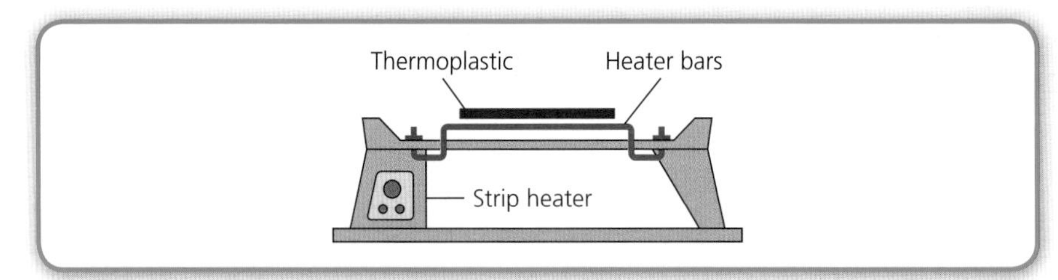

Figure 2.16: A strip heater commonly found in schools uses the same principle as commercial line-bending machines

Figure 2.17: Logos are printed onto the thermoformed casings of many products

Printing onto thermoformed products

Unlike printed products made from paper and board, when a product is thermoformed it is a complex 3D object, and products such as laptop casings cannot pass though a typical printing press. There are several methods of printing directly onto polymers that have been thermoformed, including screenprinting, hot stamping (similar to hot-foil blocking) and pad printing.

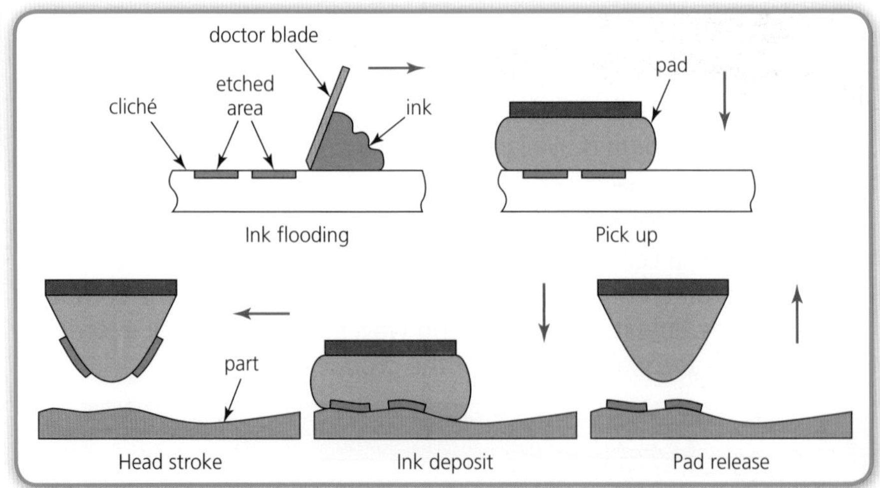

Figure 2.18: The pad-printing process

Pad printing

Pad printing allows the printing of irregularly-shaped surfaces. The image is engraved onto a printing plate known as a cliché. Ink is applied to the cliché and a doctor blade is passed across it to spread the ink evenly. A soft silicone rubber pad is then pressed against the cliché, picking up the ink and transferring it to the surface of the product.

Process	Uses	Advantages	Disadvantages	Polymers used
blow moulding	plastic bottles and containers of all sizes and shapes (e.g. fizzy drinks bottles and shampoo bottles)	• intricate shapes can be formed • can produce hollow shapes with thin walls to reduce weight and material costs • ideal for mass production – low unit cost for each moulding	• high initial set-up costs as mould expensive to develop and produce	HDPE, LDPE, PET, PP, PS, PVC
injection moulding	casings for electronic products, containers for storage and packaging	• ideal for mass production – low unit cost for each moulding for high volumes • precision moulding – high-quality surface finish or texture can be added to the mould	• high initial set-up costs as mould expensive to develop and produce	Nylon, ABS, PS, HDPE, PP
vacuum forming	chocolate box trays, yoghurt pots, blister packs, etc.	• ideal for batch production – inexpensive • relatively easy to make moulds that can be modified	• mould needs to be accurate to prevent webbing • large amounts of waste material produced	Acrylic, PS, HIPS, PVC
line bending	brochure/ menu/business card holders, shop signage	• ideal for one-off and batch production • straight bends are produced efficiently • precise temperature control over the heated area gives a neat, precise bend • set-up costs are low • computer-controlled line bending allows precise components to be formed time and again	• bends must be accurately marked before folding • acrylic can bubble if overheated • formed bend has to be allowed to cool for some time in the exact position or it springs back slightly	Acrylic, PS, HIPS, PVC

Table 2.5: Advantages and disadvantages of thermoforming techniques

ResultsPlus
Build Better Answers

Describe how line bending is suitable for one-off production. (3 marks)

 **Basic answers (0-1 marks)**
Give one characteristic of producing a product using line bending, with no justification as to why it is suitable for one-off production.

 **Good answers (2 marks)**
Relate one characteristic of line bending, with its suitability for one-off production.

 **Excellent answers (3 marks)**
Achieve full marks by fully justifying the reasons why line bending is a suitable process for producing a one-off product.

Support Activity

Make a list of products that are made using vacuum-forming and line-bending processes. Collect images of these products for your revision notes, so you can give examples in an exam.

Stretch Activity

Explain why thermoformed products are relatively inexpensive to produce in large quantities.

Joining techniques

Objectives

- **Explain** the reasons for selecting the most suitable adhesive for joining like and unlike materials.
- **Understand** the Control of Substances Hazardous to Health (COSHH) regulations when using adhesives.

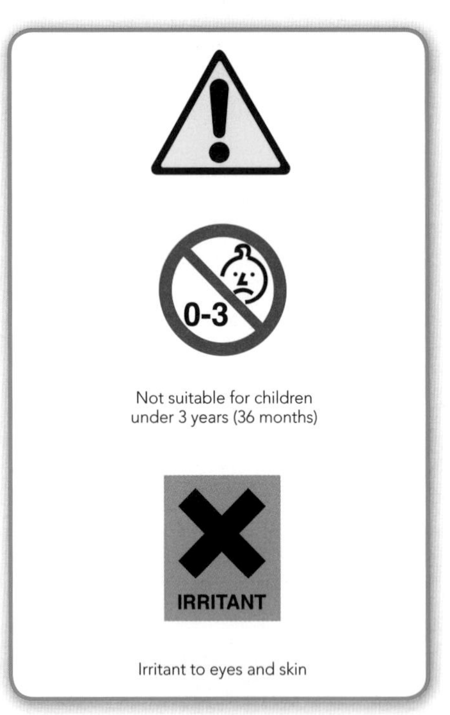

Figure 2.19: Common warning symbols found on adhesives

ResultsPlus
Watch out!

As well as being asked about a suitable adhesive for joining different materials, you may be asked to carry out a risk assessment for using an adhesive. See 'Health and safety' on pages 58–59.

Joining techniques can be divided into three main groups:

- permanent – once made they cannot be reversed without causing damage, e.g. welding metals
- temporary – although not always designed to be taken apart, they can be disassembled if needed without causing damage, e.g. flat-pack furniture
- adhesives – substances for bonding two or more materials together using a chemical reaction, e.g. using PVA glue to join MDF.

Adhesives

There is a wide range of adhesives available, from general-purpose adhesives that can join a variety of materials to specialist adhesives designed to join specific materials only. Some adhesives can join 'like' materials only: for example, Tensol® cement is designed to join acrylic to acrylic; other adhesives can join 'unlike' materials: for example, epoxy resin can join wood to metal.

Table 2.6 shows four of the most popular types of adhesive that you could use when making models.

Preparation

Adhesives require clean and thoroughly prepared surfaces in order for the joint to fully bond.

- As polymers often have a smooth, shiny surface finish, they should be cleaned and roughened with an abrasive paper before bonding.
- Metal surfaces should be degreased and roughened with an abrasive paper.
- Woods are porous so the adhesive will soak into the surface. Once the adhesive has been applied to the cleaned and prepared surface, it then has to cure (harden). It is important that the two surfaces are held in place securely so that they do not slip as this process can take several minutes.

Warning symbols

Warning symbols are placed on products to provide health and safety information for the consumer. Many warning symbols appear on the packaging of adhesives, along with additional safety instructions that outline any potential risks to users.

Support Activity

Collect off-cuts of materials from your workshops and design studios.
As a class, experiment with various adhesives to join different pieces of scrap material. Make a table showing which adhesives join like and unlike materials the best.

Adhesive	Uses	Advantages	Disadvantages
epoxy resin	most materials including expanded polystyrene (EPS)	• high-performance adhesive giving high-strength bonds • chemical reaction hardens immediately • versatile – can be made flexible or rigid, transparent or opaque/coloured, rapid or slow setting • excellent heat and chemical resistance	• reaches full strength after a few days • expensive • often requires manual mixing of resin and hardener, which can be messy
polystyrene cement	polystyrene (PS) and high-impact polystyrene (HIPS) for vacuum forming Note: should not be used on expanded polystyrene (EPS) as it dissolves	• strong bond: melts surface of pieces to be joined and causes them to weld together • able to use a brush to apply (water-like consistency) and absorbed into joint by capillary action	• relatively expensive • solvent-based so contains harmful volatile organic compounds (VOCs)
Tensol® cement	acrylic	• produces high-strength bonds to acrylic sheet • clear adhesive – if acrylic is glued correctly you will not see the join	• needs to be clamped together for 24 hours to give a permanent joint • solvent-based so contains harmful VOCs
polyvinyl acetate (PVA)	woods and porous materials such as Styrofoam™	• gives a strong joint • dries clear • relatively inexpensive	• surfaces need to be securely clamped together for long periods for PVA to harden • most brands not waterproof

Table 2.6: Adhesives for joining like and unlike materials

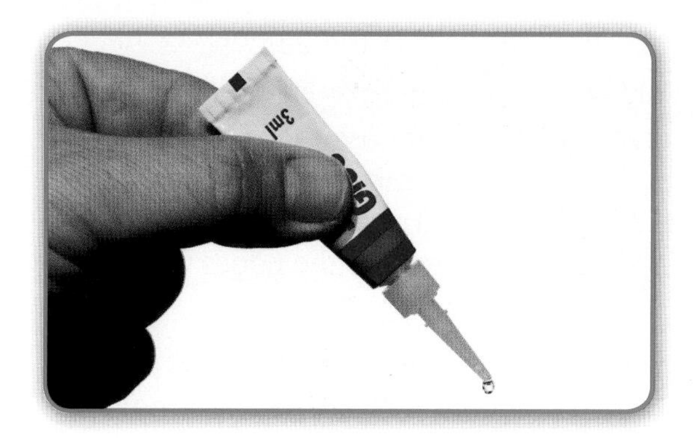

Stretch Activity

Use the internet to research manufacturers' safety instructions for the adhesives covered in this section. For example, one common manufacturer of epoxy resin is Araldite.

Finishing techniques

Objectives

- **Describe** the laminating, varnishing and hot-foil blocking finishing processes for paper and board.
- **Explain** the reasons for selecting these finishing processes for enhancing the format of paper and board.

Finishes are applied to materials in order to protect the surface and provide an improved visual appearance. For example, MDF can be spray-painted to provide a high-quality gloss finish that looks like a polymer.

There are a number of finishing processes that can be used to improve the performance, quality and aesthetic and functional properties of paper and board.

Laminating

Laminating involves applying a transparent plastic film to the surface of paper and board. Lamination has a wide range of uses across the whole spectrum of printed products due to its properties of good gloss and strength and the advantage of being low in cost.

In commercial 'heated roller' laminating, a polypropylene (PP) film is glued to the paper as it is fed through a heating wedge under high pressure. Heating the glue before applying the film to the paper or board makes the application of the film faster. As the adhesive materials are non-adhesive until exposed to heat, they are much easier to handle. The glue is solid at room temperature, so lamination of this type is less likely to shift or warp after it has been applied.

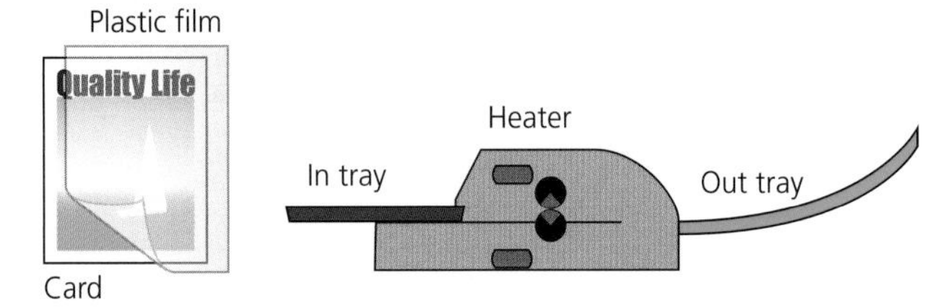

Figure 2.20: The 'heated roller' lamination process

Support Activity

1 Collect examples of all three finishing processes. Look out for examples that use more than one process in one product.

2 What other finishing techniques can be used to enhance the format of paper and board?

Stretch Activity

Contact a local printer and ask for two quotes for printing a graphic product you have designed:

- full-colour printing using a commercial printing process such as offset lithography
- as above but with a section of hot-foil blocking.

Discuss with the printer whether hot-foil blocking is worth the added cost.

Varnishing

Varnish is applied to paper and board to give it a high-gloss finish, as on the pages of glossy magazines, to give the paper a quality feel. A fine varnish is sprayed onto the surface of the paper or board. Once dry, this gives a gloss finish, which helps to protect the printing underneath.

The varnishing process can only take place after the colour printing is completed, as the oil- or water-based varnishes used take at least two hours to dry. This is a major disadvantage, as other finishing processes have to be delayed until the varnish dries. For example, a brochure cannot be collated, folded and bound too early or the pages will stick to one another.

Ultra-violet (UV) varnishing can be used to speed up the drying process. Special varnishes dry almost straight away if they are exposed to ultra-violet light. The varnish is sprayed onto the paper in the same way as other varnishes. However, after spraying the paper passes under UV lights, which dry the varnish almost instantaneously. Printed materials can move quickly onto other finishing processes. One disadvantage is that this type of machinery is expensive to purchase. However, it produces the ultimate in gloss finishes to paper and board.

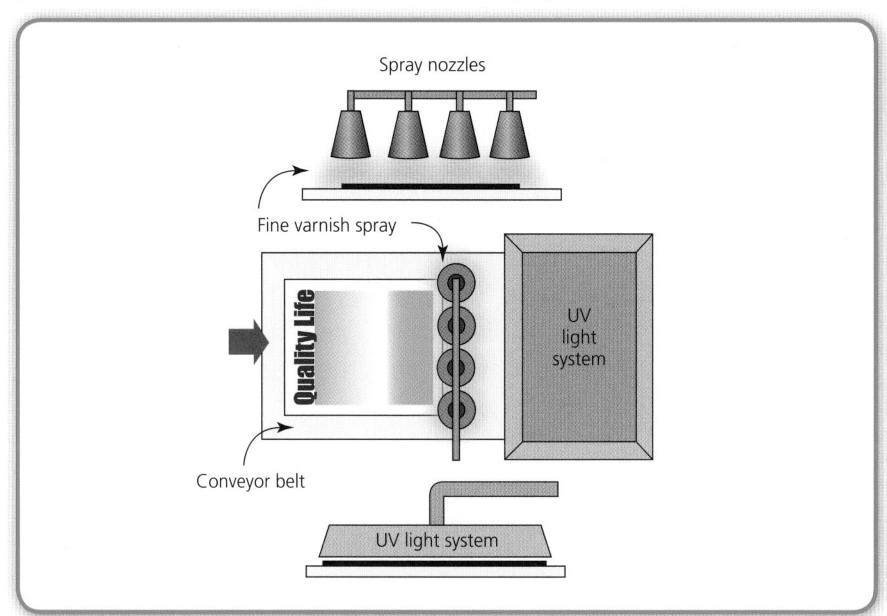

Figure 2.21: Ultra-violet varnishing process

Varnish can also be applied by the screenprinting process – this is particularly used for spot varnishing. Spot varnishing applies UV varnish to selected areas of a printed image to enhance product impact or form part of the graphic design.

Hot-foil blocking

Hot-foil blocking, also known as foil blocking or hot-foil printing, is used to produce true 'reflective metal' printing, and other effects impossible with normal metallic printing inks. Hot-foil blocking can be used to enhance and add value to conventionally printed materials.

In the hot-foil blocking process, a foil coating is transferred to paper or board by means of a heated die. A roll of foil with a polyester backing sheet is continuously fed over the paper or board, and a heated die presses the foil onto its surface.

This visual impact comes at a cost, so designers must weigh up whether the cost is worth the added visual appeal.

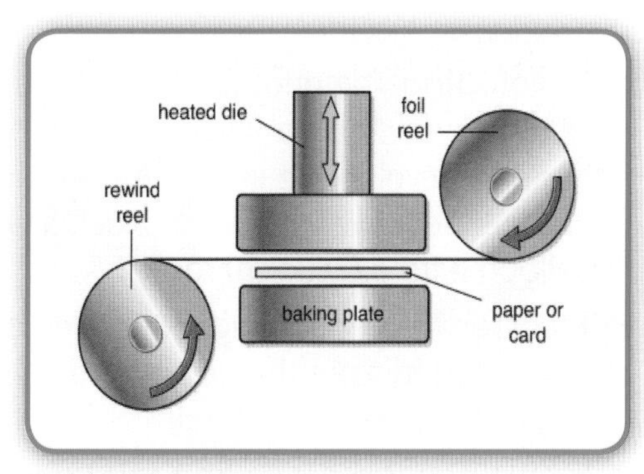

Figure 2.23: The hot-foil blocking process

Figure 2.22: An example of how UV spot varnishing can be used to enhance product impact

Apply it!

Encapsulation is one form of lamination available in most schools. Here the paper is sealed in a polythene pouch. This can be used to give paper greater strength and a high-gloss finish to many graphic products, such as menus and business cards.

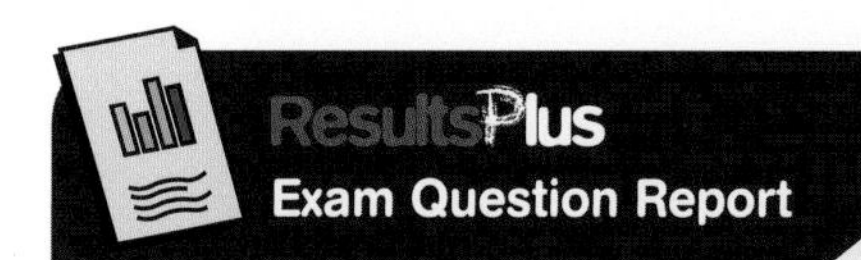

ResultsPlus
Exam Question Report

Explain two reasons why a restaurant menu would be laminated. (4 marks)

How students answered

Some students only stated one appropriate reason, with no justification.

5% 0-1 marks

Most students were able to include two valid reasons why a menu would be laminated. However, they only justified one reason in sufficient detail.

56% 2-3 marks

Many students correctly identified two properties of laminated card, such as strength/durability and visual impact/gloss finish, fully justifying each as to its suitability for a restaurant menu.

39% 4 marks

Printing processes

Objectives

- **Describe** the photocopying, offset lithography, flexography, gravure and screenprinting processes.
- **Explain** the reasons for selecting specific printing processes for creating graphic products.

Commercial printing processes are distinguished by the method of image transfer used. Depending on the process, the printed image is transferred to the paper either directly or indirectly.

- In *direct* printing, the image is transferred directly from the plate cylinder (or image carrier) to the paper: for example, gravure, flexography, screenprinting and photocopying processes.
- In *indirect*, or offset, printing, the image is first transferred from the plate cylinder to the blanket cylinder, and then to the paper: for example, offset lithography, which is the most widely used commercial printing process.

Photocopying

Photocopying is a widely available printing process, both in schools and through copy centres on the high street. Full-colour copying using expensive digital photocopiers is now commonplace. However, even the simplest black and white copying process has a number of stages to producing an image.

Stages involved in photocopying

- **Charging** Inside the photocopier is a cylindrical drum that is electrostatically charged (think of rubbing a balloon to make a static charge). The drum has a coating of a photoconductive material that will conduct electricity when exposed to light.
- **Exposure** An intense beam of light illuminates the original document, and the white areas of this document reflect the light onto the surface of the photoconductive drum. The areas of the drum that are exposed to light (those areas that correspond to white areas of the original document) become conductive; the areas of the drum not exposed to light (those areas that correspond to black portions of the original document) remain negatively charged. The result is an electrical image on the surface of the drum.
- **Developing** The photocopier holds a fine black powder known as toner, which is positively charged. When toner is applied to the drum to develop the image, it is attracted and sticks to the areas that are negatively charged (black areas), just as paper sticks to a toy balloon with a static charge.
- **Transfer** The toner image produced on the surface of the drum is transferred from the drum onto a piece of paper using a higher negative charge than the drum.
- **Fusing** The toner is melted and bonded to the paper by heat and pressure rollers.
- **Cleaning** The drum is wiped clean with a rubber blade and completely discharged by light.

Offset lithography

Lithography works on the principle that oil and water do not mix, but repel each other. Modern offset lithography is used to produce posters, books, newspapers, packaging, credit cards, decorated CDs – just about any flat-surfaced, mass-produced item with print on it.

The development of digital imaging has enabled print shops to produce printing plates directly from a computer using direct laser imaging; this is known as computer-to-plate (CTP).

Stages of the offset lithography process

- The printing plate, made from a flexible aluminium or polymer, is fixed to the plate cylinder on the printing press. The printing plate carries the design, such as text and photographs on a magazine page, formed using an oil-based emulsion.
- Rollers apply water, which covers the blank portions of the printing plate but is repelled by the emulsion of the design area.
- Ink, applied by other rollers, is repelled by the water and only sticks to the emulsion of the design.
- The printing plate then rolls against a drum covered with a rubber blanket (blanket cylinder), which squeezes away the water and picks up the ink.
- The paper rolls across the blanket cylinder, which transfers the image to the paper.

Because the image is first transferred, or offset, to the blanket cylinder, this reproduction method is known as offset lithography or offset printing.

Offset lithographic presses involve multiple print units, each containing one printing plate for the four 'process colours' of cyan (blue), *m*agenta (deep pink), *y*ellow and blac*k* (CMYK). They are capable of printing multi-colour images on both sides of the sheet at the same time and at very high speeds.

Some presses can accommodate continuous rolls of paper or 'webs; these presses are called web presses.

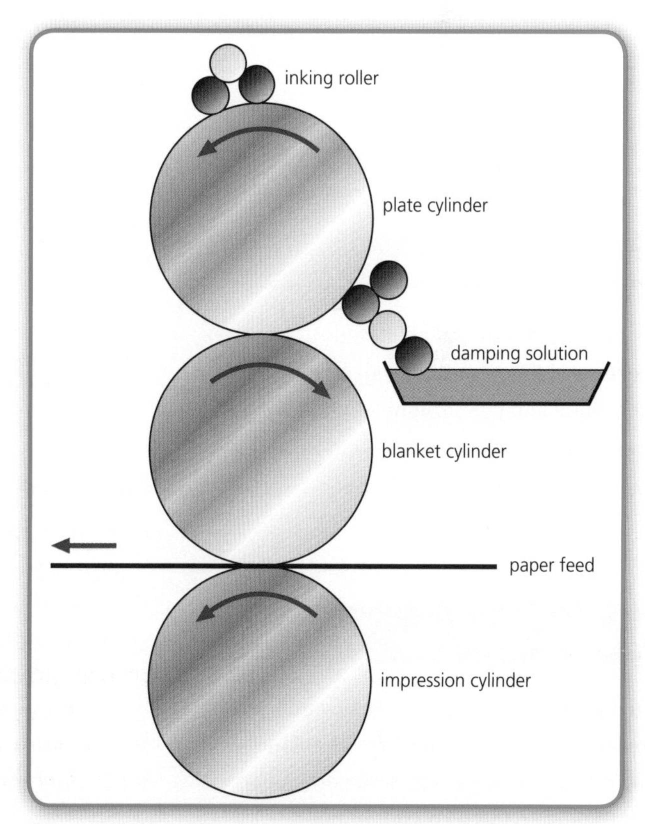

Figure 2.24: The offset lithographic printing process

Flexography

Flexography is the main printing process used to commercially print packaging materials including cartonboard containers, plastic bags and chocolate bar wrappers. Flexography uses a relief-type printing plate with raised images; only the raised images come into contact with the paper during printing. Printing plates are made of a flexible material, such as plastic, rubber or UV-sensitive polymer (photopolymer), so that it can be attached to the plate cylinder for ink application.

Flexographic presses have a plate cylinder, a metering cylinder (known as the anilox roll) that applies ink to the plate, and an ink pan. Some presses use a third roller as a fountain roll, and some have a doctor blade for improved ink distribution.

In the flexographic printing process, the paper is fed into the press from a roll. The image is printed as the paper is pulled through a series of print units. Each print unit is printed with a single process colour (CMYK). As with gravure and lithographic printing, the various tones and shading are achieved by overlaying the four basic shades of ink.

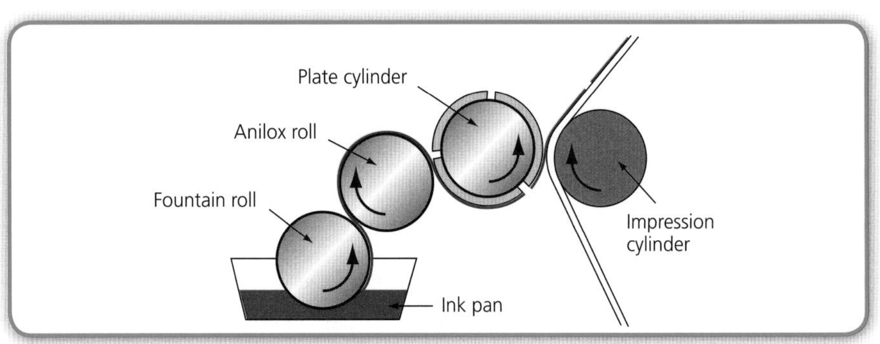

Figure 2.25: The flexographic printing process

Gravure

Rotogravure (gravure for short) is a type of printing process where the image is engraved onto a copper-plate cylinder. Gravure, like offset and flexography, uses a rotary printing press and the vast majority of presses print on reels of paper, rather than sheets of paper. Rotary gravure presses are the fastest and widest presses used commercially, and they can print everything from narrow labels to 12-feet-wide rolls of vinyl flooring. In-line finishing operations carried out on press such as saddle-wire stitching for magazines are also possible with gravure.

On the gravure plate cylinder, the engraved image comprises of small, recessed cells that act as tiny ink wells. The depth and size of these wells control the amount of ink that gets transferred to the paper.

A rotogravure printing press has one printing unit for each of the four process colours (CMYK). There are five basic components in each colour unit: an engraved plate cylinder, an ink fountain, a doctor blade, an impression cylinder, and a dryer.

Support Activity

Use the internet to find animated diagrams of the photocopying, offset lithography and flexography printing processes. These will help you to understand the processes better.

Stretch Activity

Find out from a local printer how much it would cost to produce 5000 full-colour copies of a graphic product that you have designed using the following printing processes:

- photocopying
- offset lithography

Why is there a difference in the two quotes?

Stages of gravure printing

- The plate cylinder is partially immersed in the ink fountain, filling the recessed cells. As the cylinder rotates, it draws ink out of the fountain with it.
- The doctor blade scrapes the cylinder before it makes contact with the paper, removing ink from the non-printing (non-recessed) areas.
- The paper passes between the impression cylinder and the plate cylinder under pressure. Here, the ink is transferred from the recessed cells to the paper. The purpose of the impression cylinder is to apply force, pressing the paper onto the plate cylinder, ensuring even and maximum coverage of the ink.
- The paper passes through a dryer because it must be completely dry before going through the next colour unit and absorbing another coat of ink.

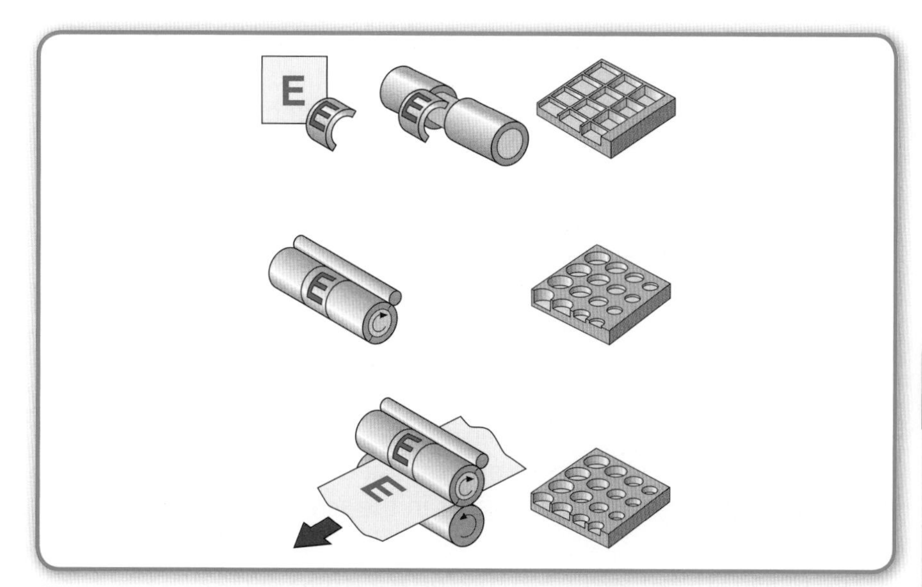

Figure 2.26: The rotogravure printing process

Explain **two** benefits of using offset lithography to print 5000 copies of a flyer for a new club night rather than using gravure. (4 marks)

Basic answers (0-1 marks)
Give just one benefit, with no justification.

Good answers (2-3 marks)
Offer two appropriate benefits of using offset lithography to print the flyer, such as quality and price. However, only justify one benefit in relation to why it is better than using a photocopier.

Excellent answers (4 marks)
Correctly identify two benefits of using offset lithography to print a high-quality flyer, and justify these to show that this method is more economically viable than photocopying.

ResultsPlus
Build Better Answers

Describe the process of printing a design onto a T-shirt using screenprinting. (4 marks)

■ **Basic answers (0-1 marks)**
State one appropriate making stage with little detail of any other stage. Responses usually lack any real technical understanding of the screenprinting process.

● **Good answers (2-3 marks)**
State up to three specific stages in the screenprinting of a T-shirt, with some technical understanding demonstrated.

▲ **Excellent answers (4 marks)**
Achieve full marks when four specific stages are described with sufficient technical detail.

Support Activity

Use the internet to find animated diagrams of the gravure and screenprinting processes. These will help you to understand the processes better.

Stretch Activity

Screenprinting is one of the few commercial printing processes that can be replicated in school. Your art or textiles department may have some screens and printing inks for you to use, to have a go at this process yourself. You could produce some T-shirts, which you may want to sell as a mini-enterprise.

Screenprinting

Screenprinting is a widely used commercial printing process for producing many mass- or large batch-produced graphics, such as posters or point-of-sale display stands. The screen is made of a piece of porous, finely woven fabric (originally silk, but these days typically made of polyester or nylon) stretched over a wooden or aluminium frame. A stencil is used which blocks off areas of the screen with a non-permeable material. This stencil is a negative of the image to be printed, so the open spaces are where the ink will appear.

Screens and stencils are produced commercially using the photo-emulsion technique.

- The original image is placed on a transparent overlay. The image may be drawn or painted directly on the overlay, photocopied, or printed with a laser printer, as long as the areas to be inked are opaque.
- The overlay is placed over the emulsion-coated screen, which is then exposed to a strong light. The areas that are not opaque in the overlay allow light to reach the emulsion, which hardens and sticks to the screen.
- The screen is washed off thoroughly. The areas of emulsion that were not exposed to light dissolve and wash away, leaving a negative stencil of the image attached to the screen.

Stages of screenprinting

- The screen is placed on top of a piece of dry paper, or other material such as glass, plastic, wood or fabric.
- Ink is placed on top of the screen, and a squeegee is used to push the ink evenly into the screen openings and onto the paper. The ink passes through the open spaces in the screen onto the material below.
- The screen is lifted away, and the surface left to dry.
- If more than one colour is being printed on the same surface, the ink is allowed to dry and then the process is repeated with another screen carrying a different stencil and using different colour of ink.

The screen can be re-used after cleaning.

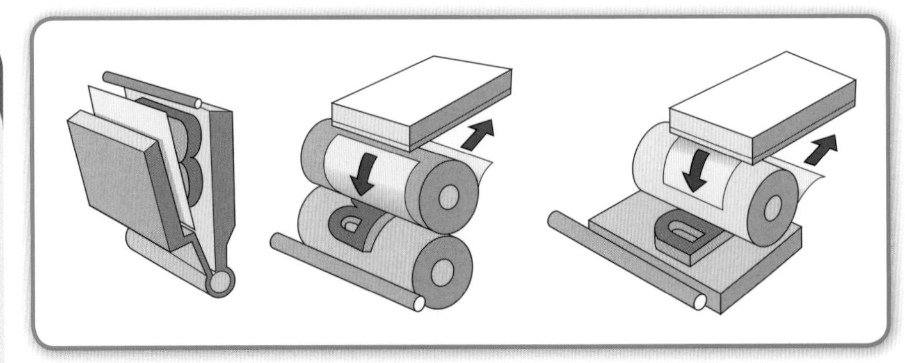

Figure 2.27: The screen-printing process

Process	Uses	Advantages	Disadvantages
photocopying	general applications such as business documents, handouts, posters	• widely available • good colour reproduction (especially on digital printers) • automatic collation and stapling of documents if required • can easily print double-sided pages • relatively low cost per copy for small batches (no set-up costs) • high printing speeds	• not cost-effective for long print runs • poor reproduction quality when toner is running low • image fades over time
offset lithography	business stationery, brochures, posters, magazines, newspapers	• good reproduction quality, especially for photographs • inexpensive printing process • can print on a wide range of papers • high printing speeds • widely available	• colour variation due to water/ink mixture • paper can stretch due to dampening • set-up costs make it uneconomic on short runs • can only be used on flat materials • requires a good-quality surface
flexography	packaging, less expensive magazines, paperbacks, newspapers	• high-speed printing process • fast-drying inks • relatively inexpensive to set up • can print on same presses as letterpress	• difficult to reproduce fine detail • colour may not be consistent • set-up costs high so would rarely be used on print runs below 500,000
gravure	high-quality art and photographic books, postage stamps, packaging, expensive magazines	• consistent colour reproduction • high-speed printing process • widest printing presses • ink dries on evaporation • variety of in-line finishing operations available • good results on lower-quality paper	• high cost of engraved printing plates and cylinders • only efficient for long print runs • image printed as 'dots' which are visible to the naked eye • very expensive set-up costs, hence only used on large print runs
screen-printing	T-shirts, posters, plastic and metal signage, point-of-sale displays, promotional items such as pens, glasses and mugs	• stencils easy to produce using photo-emulsion technique • versatile – can print on virtually any surface • economical for short, hand-produced runs • fully automatic methods capable of producing large volumes	• generally difficult to achieve fine detail (photographic screens able to reproduce fine detail) • print requires long drying times

Table 2.7: Advantages and disadvantages of different printing processes

Health and safety

Objectives

- **Understand** safe practices when designing and making graphic products.
- **Identify** hazards and risks and put in place control measures to minimise injury.

ResultsPlus
Watch out!

You may be asked to fill in blank spaces in a risk assessment table for a specific piece of equipment or process. You should familiarise yourself with safe working practices for a range of workshop equipment and processes.

The Health and Safety at Work Act 1974 states that measures should be taken, wherever possible, to safeguard the risk of injury to employees (or students). A school workshop, for example, can be a very dangerous place indeed with all of the sharp tools, equipment and machinery. Everyone using the workshop should be aware of the dangers and act in a way that reduces the risk of injury to themselves or others. Your school has a responsibility to provide you with adequate training on how to use tools, equipment and machinery, and you should never use anything that your teacher has told you not to. You are also provided with personal protective equipment (PPE) such as safety goggles and an apron and the workshop will be full of various safety signs instructing you to wear goggles on particular machines.

Health and Safety Executive risk assessments

Government guidelines for health and safety in the workplace, including schools, are laid out by the Health and Safety Executive (HSE). The HSE outlines 'Five steps to risk assessment'.

1 Identify the hazards.
2 Decide who might be harmed and how.
3 Evaluate the risks and decide on precautions.
4 Record your findings and implement them.
5 Review your assessment and update if necessary.

Hazard
potential (of risk) from a substance, machine or operation

Risk
reality (of harm from the hazard)

Control measure
action taken to minimise the risks to people

Figure 2.28: What is the difference between a hazard and a risk?

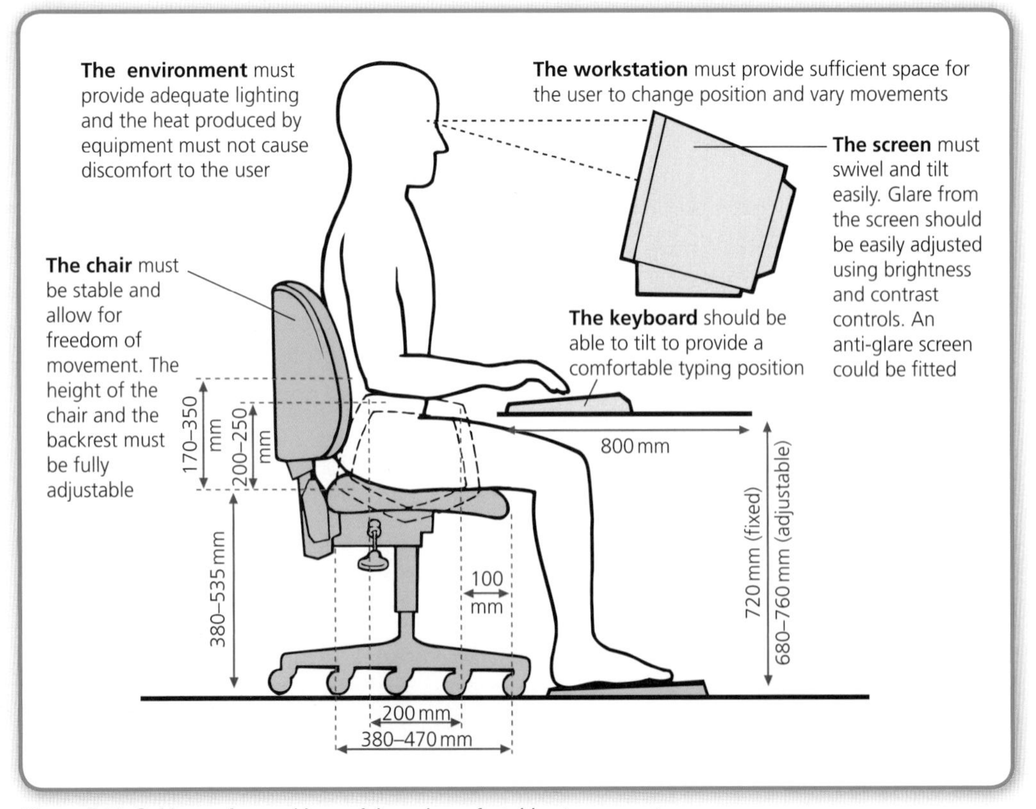

Figure 2.29: Guidance for working safely and comfortably at a computer

Using computers

Using a computer to design graphic products involves you spending large amounts of time at a workstation, looking at a screen, typing at a keyboard and using a mouse – all of which are potential hazards. See Figure 2.29 and Table 2.8.

Repetitive strain injury

Repetitive strain injury (RSI) is a medical condition affecting muscles, tendons and nerves in the arms and upper back. It occurs when muscles in these areas are kept tense for very long periods of time, due to poor posture and/or repetitive motions.

Figure 2.30: An ergonomic mouse can help prevent RSI

Hazard	Risk	Control measure
using a computer	RSI	• keyboard should tilt for comfortable typing position • use ergonomic keyboard with wrist support • use ergonomic mouse • take regular breaks to rest hands
	eye strain	• adjust screen glare using brightness/contrast controls • tilt/swivel screen to reduce reflections • take regular breaks to rest eyes

Table 2.8: Part of a risk assessment for using a computer

Support Activity

Make a risk assessment for the following hazards:

- using a knife or scalpel for cutting
- using spray adhesive for mounting graphics work.

Apply it!

Your tutor will award you marks in your Unit 1: Make Activity for health and safety. It is important that you work safely when making your final product.

Workshop practices

When making 3D models and prototypes, there are many hazards involved in using machinery, power tools and equipment. Your school or college should display a detailed risk assessment for each piece of equipment.

Hazard	Risk	Person at risk	Control measure
using a pillar drill	damage to eyes from flying debris	user/people in immediate area	• wear safety specs or goggles • brief user fully on use of machine, both general machine safety and specifics such as having guards in position • appropriate supervision by teacher or technician • ensure other students wait behind marked yellow lines or barriers when not using machine
	cuts from metal shavings	user	• work clamped securely in vice (never held in hand) to prevent work from catching and spinning • wear gloves • use a small 'stick' to remove large spiral shavings • place shavings into appropriate disposal container

Table 2.9: Part of a risk assessment for using a pillar drill

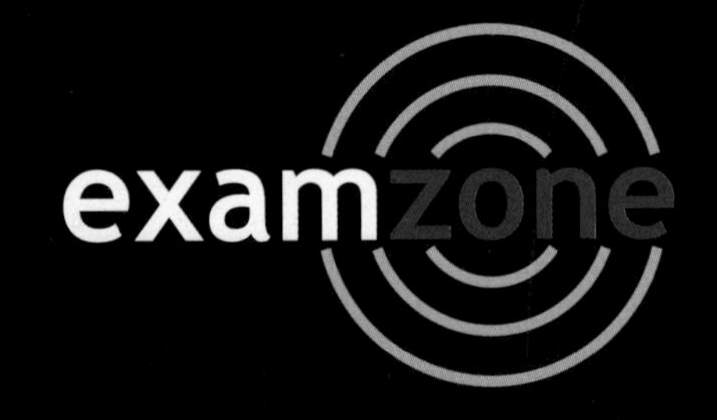

Know Zone
Chapter 2 Industrial and commercial processes

As a designer, you need to know how products are manufactured. It's quite simple - you need to be able to design products that can be made and to make products that you have designed. Many modern products are designed to be mass produced so they need to be made using processes that can produce large quantities with repeatable quality such as injection moulding. Designing with manufacturing processes in mind can be the difference between a product making a profit or failing.

You should know...

about the following about industrial and commercial processes:

Industrial and commercial process	Processes	Characteristics	Advantages and disadvantages	Uses
One-off production	✗	✓	✓	✓
Batch production	✗	✓	✓	✓
Mass production	✗	✓	✓	✓
Block modelling	✓	✗	✓	✓
Stereolithography	✓	✗	✓	✓
3D printing	✓	✗	✓	✓
Blow moulding	✓	✓	✓	✓
Injection moulding	✓	✓	✓	✓
Vacuum forming	✓	✓	✓	✓
Line bending	✓	✓	✓	✓
Adhesives	✓	✗	✓	✓
Laminating	✓	✗	✓	✓
Varnishing	✓	✗	✓	✓
Hot-foil blocking	✓	✗	✓	✓
Photocopying	✓	✗	✓	✓
Offset lithography	✓	✗	✓	✓
Flexography	✓	✗	✓	✓
Gravure	✓	✗	✓	✓
Screenprinting	✓	✗	✓	✓

Key terms

Questions relating to industrial and commercial processes will require you to know the actual process of making a product and explaining why a process is suitable for making particular products.

Property	Meaning
Processes	A description of the series of actions needed to produce a product or component. For example, knowing the stages in the injection-moulding process.
Characteristics	Recognisable features that help to identify or differentiate one process from another. For example, looking at a product made from a polymer and knowing what particular thermoforming process would be used to manufacture it.
Advantages and disadvantages	Qualities and features favourable to success or failure. For example, thermoforming can produce mass-produced products but what are the environmental implications?

Grade E-C range question: Explain **one** reason why the vacuum-forming process is suitable for manufacturing a yoghurt pot. (2 marks)

Student answer	Examiner comments	Build a better answer
Vacuum forming can be used to make plastic products. (1 mark)	Vacuum forming is a thermoforming process that can be used to produce products made from a range of polymers. This response, however, does not go on to justify why it is used for making yoghurt pots.	Vacuum forming is a suitable thermoforming process (1 mark) capable of mass-producing identical yoghurt pots. (1 mark)

Overall comment: This learner does not achieve full marks in this 'explain' question because they do not fully justify the point made. Remember, explain questions usually carry 2 marks: making a valid point (1 mark) and then justifying it (1 mark).

Grade C-A range question: Describe how a 3D printing rapid prototyping machine manufactures a model. (4 marks)

Student answer	Examiner comments	Build a better answer
The 3D printer builds up a complete model by making it out of layers (1 mark) *from a CAD file.* (1 mark)	This learner has got a good grasp of the rapid prototyping process but does not go into enough detail.	A CAD file is exported to a 3D printer (1 mark) where it is sliced into hundreds of different layers (1 mark). The 3D printer then builds up a complete model layer by layer (1 mark) by bonding layers of powder together. (1 mark)

Overall comment: Always look at the amount of marks available and the space allocated on the exam paper for your response. These are good indications of the level of detail needed.

Chapter 3 Analysing products

Objectives

- **Understand** the structure of Question 13: Analysing products.
- **Understand** how to analyse the materials and components used in a product.
- **Understand** how to analyse the manufacturing processes used in a product.
- **Understand** how to analyse the design specification of a product.

Your GCSE Graphic Products exam paper is written using a template format. The best way to explain the structure of this question is to work through a sample exam question.

First, you will be given a drawing or a picture of a graphic product, which will be clearly labelled.

Sample exam question

The drawings below show an aluminium box used to package chocolate biscuits.

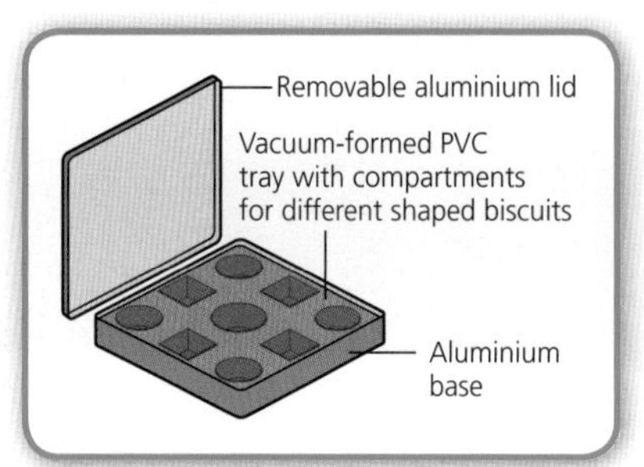

The drawing above shows the box opened.

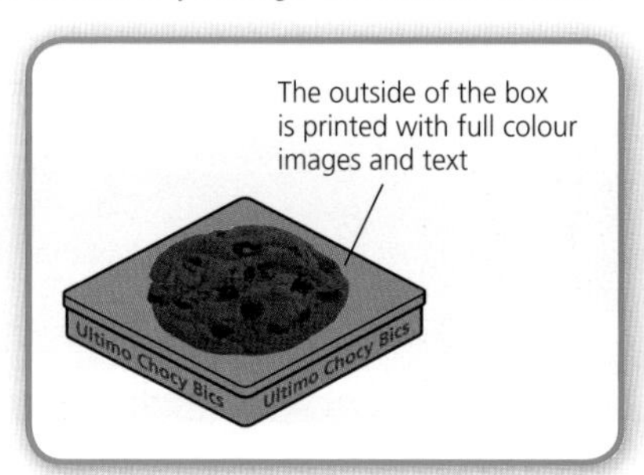

The drawing above shows the box closed, with the graphics on the outside.

(a) Give **two** properties of aluminium that make it a suitable material for packaging.

For each property, justify your answer. (4 marks)

Property 1: Aluminium is lightweight. (1 mark)

Justification: This means that transport costs can be reduced due to lighter loads. (1 mark)

Property 2: Aluminium is soft and malleable. (1 mark)

Justification: This means it can be easily formed into the shape of the tin. (1 mark)

Materials, components and manufacturing processes

The first set of questions will relate to the materials, components and manufacturing processes used in the product. When analysing a product, you should be able to identify the materials and components used in the manufacture of the product, including:

- the properties and qualities of the materials and components – what are the special features of a material? For example, glass is an inert material that will not react with the liquid contained inside it.
- the advantages and disadvantages of the materials and components – what are the benefits and drawbacks of using a specific material? For example, an advantage of glass is its high-quality appearance, whereas a disadvantage is that it is often more expensive to manufacture than polymers.

- justification of the choice of materials and components – why are certain materials used in preference to others? For example, PET is often used instead of glass because it will not shatter if dropped.

You should also be able to identify the processes involved in the manufacture of the product, including:

- the stages of the manufacturing process
- the advantages and disadvantages of the manufacturing process – what are the benefits and drawbacks of using this process? For example, injection moulding can produce relatively inexpensive products, but it has high initial set-up costs.
- justification of the choice of manufacturing process – why are certain processes used in preference to others? For example, injection moulding is used instead of blow moulding for electronic products because a thicker, more protective casing can be produced.

We know from our knowledge and understanding of materials that aluminium is a lightweight material. This property would be desirable for a packaged product, as it could be easily transported. Reductions in the weight of packaging also mean that transport costs can be cut: for example, the lightened load on a lorry would result in greater fuel efficiency.

We also know that aluminium is soft and malleable, so it can be formed into a variety of shapes, such as drinks cans. This means that the square base and lid of the product could easily be formed using aluminium.

(b) The drawing below shows the vacuum-formed PVC inner tray.

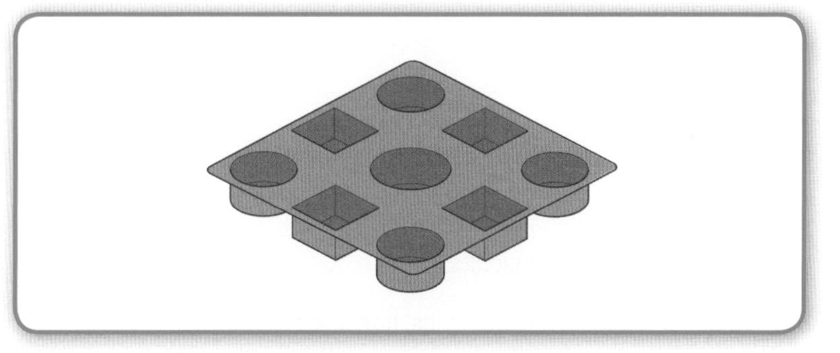

Explain **one** reason why the vacuum-forming process is suitable for manufacturing the PVC inner tray. (2 marks)

Vacuum forming is a suitable process for thermoforming PVC **(1 mark)** therefore producing identical products in large numbers. **(1 mark)**

Again, from our knowledge and understanding of manufacturing processes, we know, for example, that vacuum forming is ideal for batch and mass production of products, because it can produce intricate shapes in quantity with repeatable quality. We also know that PVC is a thermoplastic, so it can be easily thermoformed using the vacuum-forming process.

Support Activity

Some biscuit and chocolate boxes are made from polymers.

1 Answer question part (a) as though the biscuit box were made from polypropylene (PP):

(a) Give **two** properties of polypropylene that make it a suitable material for packaging. For each property, justify your answer. (4 marks)

2 Answer question part (b) as though the biscuit box were made using the injection-moulding process:

(b) Explain **two** reasons why the injection-moulding process is suitable for manufacturing the biscuit box. (4 marks)

Specification criteria

The next parts of the question will relate to the specification criteria for the product. These cover:

- **form** – why is the product shaped/styled as it is?
- **function** – what is the purpose of the product?
- **user requirements** – what qualities make the product attractive to potential users?
- **performance requirements** – what are the technical considerations that must be achieved within the product?
- **material and component requirements** – how should materials and components perform within the product?
- **scale of production and cost** – how does the design allow for scale of production and what are the considerations in determining cost?
- **sustainability** – how does the design allow for environmental considerations?

You will be asked to explain how the product meets these specification criteria.

Protection of the biscuits is a 'function' specification point. The box protects the biscuits, for example, because aluminium forms a rigid box structure. It is tough enough to take bashes and scrapes without the biscuits inside being affected.

Promoting sales of the biscuits is a 'user requirements' specification point. For example, the box has full-colour printed graphics, which means that there is total graphic coverage of the package, and customers will be able to see it from every angle when it is stacked on retailers' shelves.

(c) Two specification points for the box are that it must:

- protect the biscuits
- promote sales of the biscuits.

Explain, under the following headings, how the box meets these specification points:

(i) Protect the biscuits. (2 marks)

(ii) Promote sales of the biscuits. (2 marks)

(i) **Protect the biscuits:** The biscuits are protected by a rigid aluminium box **(1 mark)**, which is tough enough to take bashes and scrapes without the biscuits inside getting broken **(1 mark)**.

(ii) **Promote sales of the biscuits:** Graphics are printed all the way around the box. **(1 mark)** This means that customers can see the brand from all directions and easily identify it. **(1 mark)**

Stretch Activity

Use the internet to research the different ways in which biscuits and chocolates can be packaged. Look at the materials used and the manufacturing processes used to make them.

Support Activity

Use the specification criteria headings to analyse a product that you are familiar with, e.g. your MP3 player or mobile phone.

Product comparison – stretch and challenge

The last part will be an 'extended-writing' type of question, which is designed to 'stretch and challenge' the most able students. Don't worry if you are not confident in writing essays; there are a number of different ways of responding to this type of question.

You will be given a drawing or picture of a similar product and asked to compare it to the original product in terms of key specification points or other topics included in Chapter 2, such as sustainability.

(d) Packaging A and B below show two different types of packaging for biscuits.

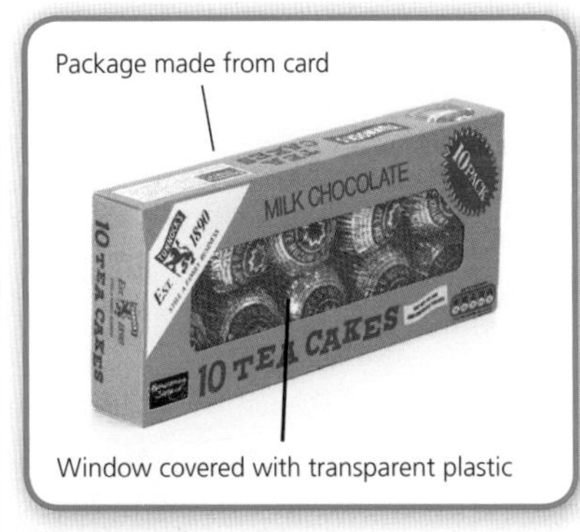

Packaging A

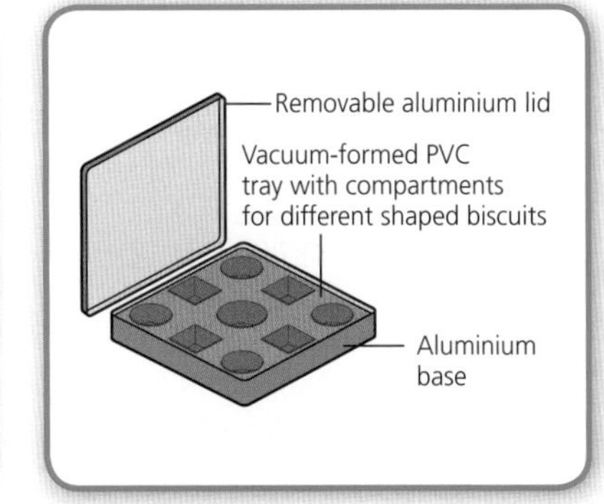

Packaging B

Evaluate packaging A compared to packaging B in terms of minimising waste production. (6 marks)

This question relates to 'minimising waste production', which in Chapter 6: Sustainability refers to the '4 Rs': reduce, reuse, recover and recycle. So you will have to compare the two different types of packaging with reference to the 4 Rs.

If writing full essays is not your style, your response could be in the form of a series of bullet points outlining the main differences between the two products. Dividing the page into two columns headed 'Packaging A' and 'Packaging B' with bullet points written under each heading would also be a clear way of presenting your answer.

Packaging A	Packaging B
• single use cartonboard package that will simply be thrown away after use (1 mark)	• aluminium tin can be reused for storage of biscuits or other things (1 mark)
• cartonboard can be recycled only when the plastic window is removed (1 mark)	• aluminium tin and PVC inner tray can be separated and recycled (1 mark)
• cartonboard requires less energy to produce (1 mark)	• aluminium takes a lot of energy to produce, from extraction of bauxite to production of alumina (1 mark)

Stretch Activity

Use the specification criteria headings to compare and contrast two similar products, e.g. two perfume bottles or two games consoles.

Know Zone
Chapter 3 Analysing products

It is important that you analyse existing graphic products to give you an insight into the work of professional designers and how they have satisfied a design brief. By doing this you will gain a greater understanding of commercial design and industrial manufacturing processes, which will in turn inform and influence your own design and make activities.

You should know...

the following about analysing a product:

- the properties and qualities of the materials and components used
- the advantages and disadvantages of the materials and components used
- how to justify the choice of materials and components used
- the advantages and disadvantages of the manufacturing processes used
- how to justify the choice of manufacturing processes used
- how the product meets the specification points given
- how to compare and contrast two similar products using the same criteria.

ResultsPlus
Maximise your marks

The picture below shows a traditional board game.

Give **two** properties of solid white board that make it a suitable material for the playing board and its box. For each property, justify your answer. (4 marks)

Learner answer	Examiner comments	Build a better answer
Property 1: *It is strong* (1 mark) **Justification:** *So the box and board will be strong* **Property 2:** *It looks good* (1 mark) **Justification:** *So the box and board will look good*	Strength is an appropriate property of solid white board but the learner does not offer a reasonable justification. 'It looks good' is given 'benefit of the doubt' as an aesthetic property. Again, the learner fails to offer a reasonable justification for the second mark.	**Property 1:** Solid white board has good strength. (1 mark) **Justification:** It enables the box to fully protect the parts inside. (1 mark) **Property 2:** Solid white board has a good quality surface finish. (1 mark) **Justification:** It enables high quality full-colour printing on the box and playing board. (1 mark)

Overall comment: This type of question is similar to an 'explain' question. As you can see, two properties need to be stated and then you need to go on to justify each of these properties. Here 'benefit of the doubt' has been used by the examiner, because the learner has made a point that is correct but could have been phrased better.

Explain how the board game is successful at meeting the following specification points:

- advertising the board game (2 marks)
- containing the playing pieces (2 marks).

Learner answer	Examiner comments	Build a better answer
Advertising the board game: *The graphics on the box are easily recognisable as Monopoly* (1 mark) **Containing the playing pieces:** *The vacuum-formed tray keeps all the pieces together* (1 mark)	The learner has given two very good reasons to match the specification points, but has not then gone on to justify each point fully.	**Advertising the board game:** The box has great visual impact (1 mark) so customers can easily tell it apart from other board games. (1 mark) **Containing the playing pieces:** The vacuum-formed tray provides individual compartments (1 mark) that keep the pieces separate and secure. (1 mark)

Overall comment: Don't forget that 'explain' questions carry 2+ marks, so you will always need to give a valid point plus justification.

Chapter 4 Designing products

Objectives

- **Understand** the assessment criteria for Question 12: Designing products.
- **Understand** how to analyse the design specification.
- **Understand** how to design ideas that address the specification criteria.

Apply it!

Use the specification criteria headings (form, function, etc.) as the basis for your specification when you write it for your Unit 1: Creative Design and Make Activities. You should be able to write a few points under each heading.

Sample exam question

A company needs to package its new 'Ultra-Torch' for sale in shops.

The torch uses an ultra-bright light emitting diode (LED), which is energy-efficient.

The picture on the right shows the torch that needs to be packaged.

Overall dimensions:

length = 20 cm

diameter = 5 cm

As the GCSE Graphic Products exam paper is written to a template format, you can be sure that Question 12 will always be the design question, year after year. The assessment of this question is also quite straightforward.

You will be given a design brief and a list of eight specification criteria. You will have to produce two different design ideas that meet the brief and fulfil the criteria.

Specification criteria

When designing a product for this question, you will be given a design brief and specification criteria which cover the following eight points:

- **form** – how should the product be shaped/styled?
- **function** – what is the purpose of the product?
- **user requirements** – what qualities would make the product attractive to potential users?
- **performance requirements** – what are the technical considerations that must be achieved within the product?
- **material and component requirements** – how should materials and components perform within the product?
- **scale of production and cost** – how will the design allow for scale of production and what are the considerations in determining cost?
- **sustainability** – how will the design allow for environmental considerations?

Design specification

The specification for the package of the torch says that it should:

- securely hold the torch inside the package
- clearly show the full length of the torch inside the package
- allow customers to easily remove the torch from the package
- be easily stacked or hung in shops for retail display
- display the brand name 'Ultra-Torch' on the outside of the package
- use graphics that make the consumer know that the product is energy-efficient
- be manufactured using appropriate materials
- be manufactured using processes suitable for mass production.

This question asks you to design two different ways of packaging a torch. First, you must analyse what is needed to address all of the specification points. Here are some questions you should ask yourself before you start designing.

- How can my designs hold the torch securely while it is in transit? For example, can internal packaging prevent the contents from moving around a box? How about a blister pack with a snug PVC blister?
- How can my designs allow the torch to be clearly visible within the package? For example, would a clear plastic window in a box work? A clear PVC blister?
- How can my designs allow customers to easily remove the torch when purchased? For example, will flaps and tucks on a box allow the package to be opened at either end? How about having two halves of a clamshell or blister pack?
- How can my designs allow the shop to either hang the product on racks or stack it on shelves for display? For example, can I include a 'euroslot' for hanging? What shape would tessellate easily for stacking?
- How can my designs display the brand name 'Ultra-Torch' on the outside of the package? For example, can I include areas where the brand name could be printed on the box and backing card?
- How can my designs imply energy-efficiency on the package? For example, what energy saving slogan could be printed on the box? How about using 'green' graphics to show 'green' credentials?
- Which materials are appropriate for my designs? For example, PVC can be used for the blister as it can be vacuum-formed easily and is transparent; folding boxboard can be used for a box as it can be easily scored, folded and printed on.
- Which mass-manufacturing processes are appropriate for my designs? For example, vacuum forming can produce a PVC blister, and die cutting can produce the net for a box. Which printing processes could be used for the graphics?

Application of knowledge and understanding

When designing your product, you should apply your knowledge and understanding of a wide range of materials, components and manufacturing processes to each design idea. Your choice of appropriate materials, components and manufacturing processes should be based on their properties, advantages and disadvantages. You should show these and make sure that your choices are justified throughout your sketches and annotation.

You may want to use a scrap piece of paper to 'brainstorm' your ideas. However, do not attach these additional pieces of paper to your exam as they will not be marked by an examiner. Also, avoid scribbling on the exam paper itself as this may detract from your actual answer.

Support Activity

Practice analysing a design specification by attempting Question 12s from the sample assessment materials (SAMs) or past exam papers.

Stretch Activity

Set yourself some simple design tasks and sketch two completely different design ideas that are fully annotated.

Designing skills

When designing a product in this exam, you should be able to respond creatively to the design brief and specification criteria that are given, including:

- clear communication of your design ideas using notes and sketches
- annotation that relates to the original specification criteria
- application of your knowledge and understanding of materials, components and manufacturing processes.

Eight marks are available for each design idea, and there are eight specification points. So to achieve full marks, you will need to address every specification point for each of your designs.

Here is one learner's response to the specification criteria set earlier (see page 66).

Design idea 1 (8 marks)

Design idea 1 focuses on a packaging net concept. The learner has sketched several clear drawings to illustrate their design idea fully (see Figure 4.1). The triangular package is drawn in 3D to show how it would look assembled with the graphics in place, while smaller 'thumbnail' sketches show more technical details, such as the 2D net involved and internal supports. The annotation and drawings clearly address all eight specification points, as this design:

- holds the torch securely while in transit by using two internal supports within the package to prevent contents from moving (1 mark)
- allows the torch to be clearly visible within the package by use of an acetate/clear plastic window (1 mark)
- allows customers to remove the torch easily once purchased using flaps and tucks indicated on a net to open either end of the package (1 mark)
- allows the shop to stack the product on shelves for display because its triangular shape tessellates easily (1 mark)
- clearly displays the brand name 'Ultra-Torch' on the outside of the package (1 mark)

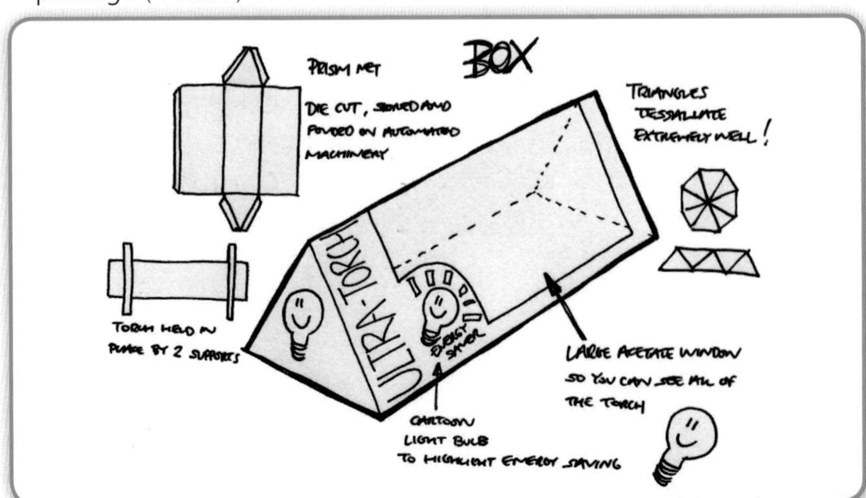

Figure 4.1: This learner's first idea is based on a triangular packaging net

- implies energy-efficiency on the package by using a 'happy' light-bulb character with the slogan 'energy saver' (1 mark)
- indicates one material appropriate to the design – acetate for the window (a suitable cartonboard could also have been mentioned to secure the mark) (1 mark)
- indicates one mass-manufacturing process appropriate to the design – die cutting of the net (a suitable printing process could also have been mentioned to secure the mark) (1 mark).

Design idea 2 (8 marks)

Design idea 2 **must** be a totally different concept to your first idea. This learner's second idea focuses on a blister pack concept, which is clearly different to a packaging net. The learner has sketched the blister pack intact at first, and then with the torch being removed to illustrate how easy it would be to open the blister (see Figure 4.2). Again, the annotation and drawings clearly address all eight specification points, as this design:

- holds the torch securely while in transit by using a vacuum-formed PVC blister, which fits snugly around it (1 mark)
- allows the torch to be clearly visible within the package using a clear PVC blister (1 mark)
- allows customers to remove the torch easily when purchased, using a blister that can be easily ripped off the backing card (1 mark)
- allows the shop to hang the torch on racks for display using a euroslot (1 mark)
- clearly displays the brand name 'Ultra-Torch' on the backing card (1 mark)
- implies energy-efficiency on the backing card by using a 'happy' sun character which shows its 'green' credentials (1 mark)
- indicates two materials that are appropriate to the design – PVC for the blister and card backing (please note that you can't achieve two marks for naming two materials) (1 mark)
- This design indicates two mass-manufacturing processes appropriate to the design – vacuum-forming for a blister and offset lithography for printing the graphics (again, you can't achieve two marks for naming two manufacturing processes) (1 mark).

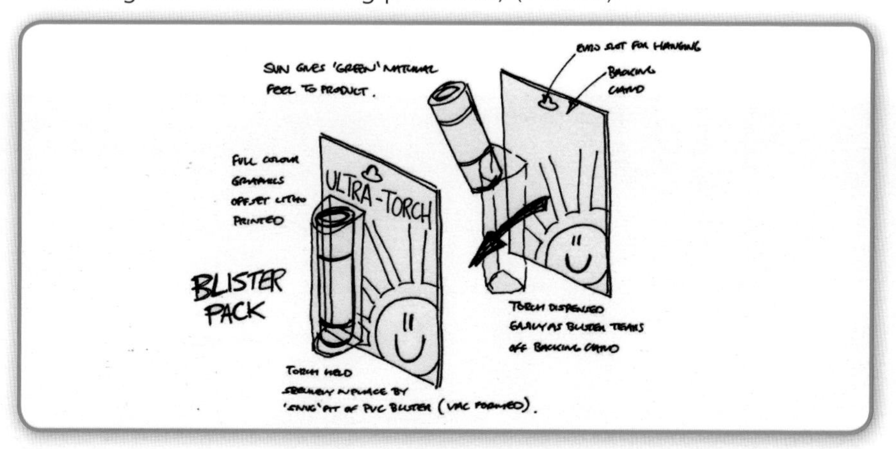

Figure 4.2: This learner's second idea is a totally different concept to the first by using a blister pack

Your exam paper will be scanned into a computer before it is marked, so your drawings must be clear and easy to read. You do not have to use colour and/or shading at all as there are no marks available for the presentation of your ideas. Clear line drawings using an HB pencil or a black fine-liner pen would be ideal.

Support Activity

Practice your designing skills by attempting Question 12s from the sample assessment materials (SAMs) or past exam papers.

Stretch Activity

Set yourself some simple design tasks and sketch two completely different design ideas that are fully annotated.

Know Zone
Chapter 4 Designing products

Generating ideas is an extremely important part of the creative design process. This is the stage where you can demonstrate your individual flair and talent by creating alternative ideas that fulfil a set of specification criteria.

You should know...

the following about designing a product:

- how to respond creatively to a design brief and specification criteria
- how to clearly communicate your ideas using notes and sketches
- how to use annotation that relates back to specification criteria
- how to apply your knowledge and understanding of materials, components and manufacturing processes to your ideas.

ResultsPlus
Maximise your marks

A manufacturer requires a point-of-sale (POS) display for the packs of sweets. The POS will be placed on a counter of a shop. The drawing below shows one pack of sweets.

Each pack of sweets is 50 mm long with a diameter of 20 mm.

Design specification

The specification for the point-of-sale display state that it should:

1. be freestanding
2. hold at least 10 packs of sweets
3. allow customers to remove the sweet.
4. allow the point-of-sale display to be refilled.
5. display the name of the product, 'Solo'
6. display a graphic or text to encourage sales
7. be manufactured using appropriate materials
8. be manufactured using processes suitable for batch production.

Learner answer	Examiner comments	Build a better answer
Design idea 1 (7 mark) 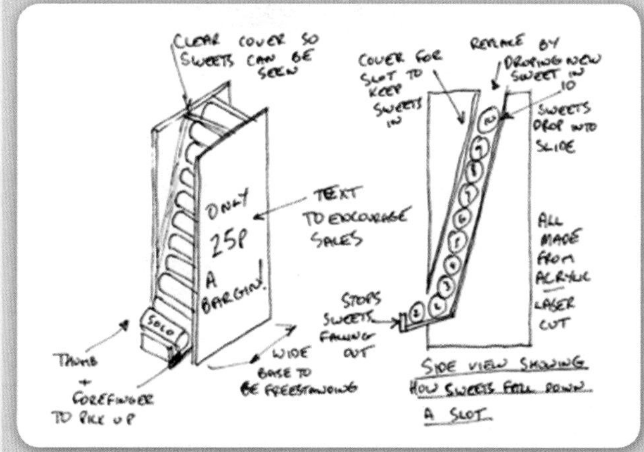	This learner has presented their first design idea well. The sketches are clear and well annotated. Annotation relates to specification criteria.	The learner has not answered specification point 5: display the name of the product, 'Solo'. One of the packs of sweets is labelled, but the point-of-sale display does not carry the brand name.
Design idea 2 (7 mark) 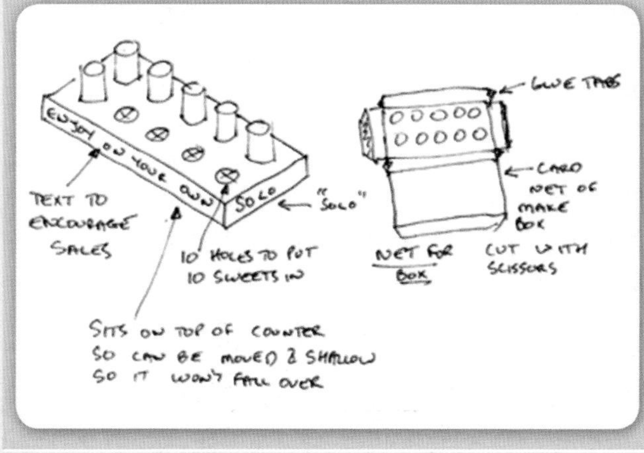	A good attempt – this second idea is completely different to the first. Although the idea is not as practical as the first, it still fulfils most of the specification criteria. Where annotation is lacking, the examiner has to make a decision based on the sketch.	'Cut with scissors' is not really a suitable batch-production process. A suitable process here would be die-cutting.

Overall comment: Make sure that you address all eight specification points given. Use the specification as a checklist for both of your designs.

Chapter 5 Technology
Information and communications technology

Objectives

- **Explain** the advantages and disadvantages of using email for communication.
- **Understand** how the internet is used for marketing and sales of products.
- **Describe** the electronic point of sale (EPOS) system and how it is used for marketing and sales of products.

Information and communications technology (ICT) has had a dramatic effect on our lives. We can now communicate quickly over vast distances using email, and shop online without leaving home. Modern businesses have had to embrace ICT, from professional websites to market products to EPOS systems to monitor sales and stock levels.

Email

Email is the simplest form of electronic communication. Email has relatively low reach and range when used for messaging and transferring documents, but is invaluable in rapid communications between designers, manufacturers, retailers and consumers because it is easy to use and access. However, with email there are issues of security and privacy, and limitations on the size of attachments.

To most people, email is now the preferred way of communicating, especially with the increased use of multimedia phones such as the iPhone and social networking sites such as Facebook and Twitter.

Advantages	Disadvantages
• quick, easy and convenient communication around the world • widespread usage by anyone connected to the internet • email exchange can be saved as a dated record of correspondence • documents can be attached, saved and edited easily • social networking from mobile devices helps people keep in touch regularly	• impersonal, and some messages can be misinterpreted • dealing with flow of messages into inbox takes time • 'spamming' of unsolicited commercial emails, often with inappropriate content • privacy and security issues, as messages can be intercepted • limitations on size of attachments

Table 5.1: Advantages and disadvantages of email

Internet marketing and sales

The development of the internet as a means of competing in a global marketplace has revolutionised the marketing and sales of products and services. Through the global networking of computers, the internet gives anyone with a computer access to a wealth of information and entertainment.

ResultsPlus
Build Better Answers

Discuss the use of ICT in the design and development, manufacture, sales and marketing of products. (6 marks)

Basic answers (0-2 marks)
State up to two advantages of using ICT.

Good answers (3-4 marks)
State up to four advantages of using ICT.

Excellent answers (5-6 marks)
Achieve full marks when both advantages and disadvantages of using ICT are discussed.

The dramatic rise in e-commerce has led to virtual communities forming, which businesses are eager to explore. These new markets open up a whole range of new possibilities for innovative marketing techniques, as you can tell what someone is like by the sorts of choices they make, and the sorts of communities they join. Marketing can therefore, be 'tailor-made' to suit these markets. For example, a pop-up advert can be placed on a person's social networking site to fit with the sort of information that they have put into their user profile.

Advantages	Disadvantages
To manufacturers and retailers: • global reach and access to markets in other countries, and an increased customer base • higher global company profile • faster processing of orders, resulting in cost savings and reduced overheads • detailed knowledge of user preferences and market trends by tracking sales • reduced sales force and less need for retail outlets • less expensive than traditional media such as TV • new marketing tactics can target specific groups To the consumer: • access to a wide range of products and services • product information to inform buying decisions • online discounts and savings through price comparison websites • convenience of home shopping and delivery	• security concerns regarding input of personal bank details when purchasing goods • personal information can be shared with other companies without customer permission • risk of 'pop-up' adverts spreading viruses on computer system • difficult to find websites without exact details, resulting in a need for other expensive marketing methods such as magazine adverts or searches sponsored by the internet service provider • slow internet connection can cause difficulties in accessing information • difficulty in navigating complicated web pages • does not allow 'hands on' experience of a product such as touch, taste and fit • access to inappropriate material • spread of 'junk' mail and threat of computer viruses

Table 5.2: Advantages and disadvantages of internet marketing and sales

Support Activity

1 Why do some people buy products over the internet rather than in shops? Explain your answer.

2 Look at a sales website you know. Does it make you want to buy the products? How could it be more 'user friendly'?

Stretch Activity

Think of how you use email and the internet. What are your views on the benefits and drawbacks of email and the internet in your day-to-day life?

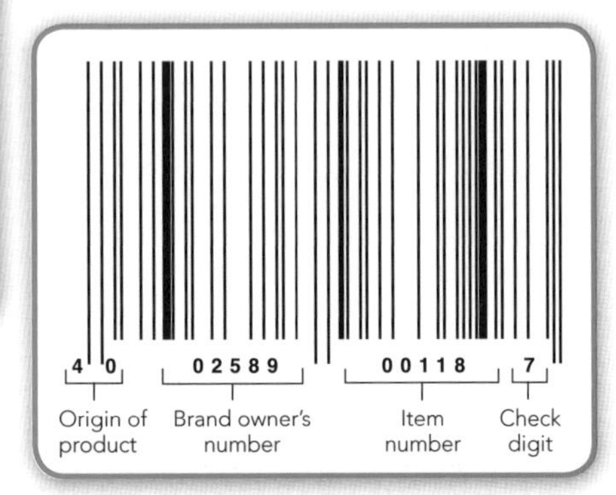

Figure 5.3: A standard 13-digit barcode

Figure 5.4: 2D barcodes, or data matrix codes, contain more information because they use both the width and height of the lines

Electronic point of sale

Information is at the centre of any business and, if used properly, ensures that the business stays one step ahead of the competition. By using electronic point of sale (EPOS) systems, a business can supply and deliver its products and services faster, by reducing the time between the placing of an order and the delivery of the product.

The use of barcodes

Each product can be electronically identified using its unique barcode. When passed over a barcode reader or scanner, the barcode is 'read' by a laser or light beam. Think of the checkout assistants at your local supermarket, and the familiar 'beep' sound as your shopping is scanned. The beam passes over the barcode and reflects back onto a photoelectric cell. The bars can be detected because they reflect less light than the background on which they are printed.

Each product has its own unique number. On the 13-digit barcode in **Figure 5.3**, the first two numbers tell you where the product was made; the next five digits are the brand owner's number; and the next five numbers are specific to the type of product (given by the manufacturer). The final digit is the check digit, which confirms that the whole number has been scanned correctly.

It is important to note that information regarding the actual price of the product is not contained on the barcode. Instead, the scanner, at a supermarket checkout, for example, transmits the product code number to an in-store computer; this computer relays the product's description and price check back to the checkout. This price is then displayed electronically on the till and printed on the receipt. The computer then deducts the item purchased from the store's stock list, so that they can be re-ordered when stock is low.

Data matrix codes, also known as 2D barcodes, are visual codes that can be read by machines with vision systems (see the example in Figure 5.4). Data matrix codes are being used more and more, as manufacturers want to keep closer track of their products. Data matrix codes mean that batch or serial numbers can be permanently marked onto components, which is useful for tracking faulty batches and identifying counterfeit parts.

Advantages of EPOS

EPOS provides manufacturers with:

- a full and immediate account of the sales of a company's products
- data that can be input into spreadsheets for sales analysis
- the means to monitor the performance of all product lines, which is particularly important with mass production as it allows a company to react quickly to demand

Support Activity

Look at Figure 5.5. Copy the diagram but insert your own example to illustrate the process: for example, a new pair of Nike trainers. Where is the product manufactured (it may not be in this country)? Where did you buy the product? How much did it cost?

- accurate information for identifying consumer buying trends when making marketing decisions
- a full and responsive stock control system by providing real-time stock updates
- a system that ensures sufficient stock is available to meet customer needs without overstocking.

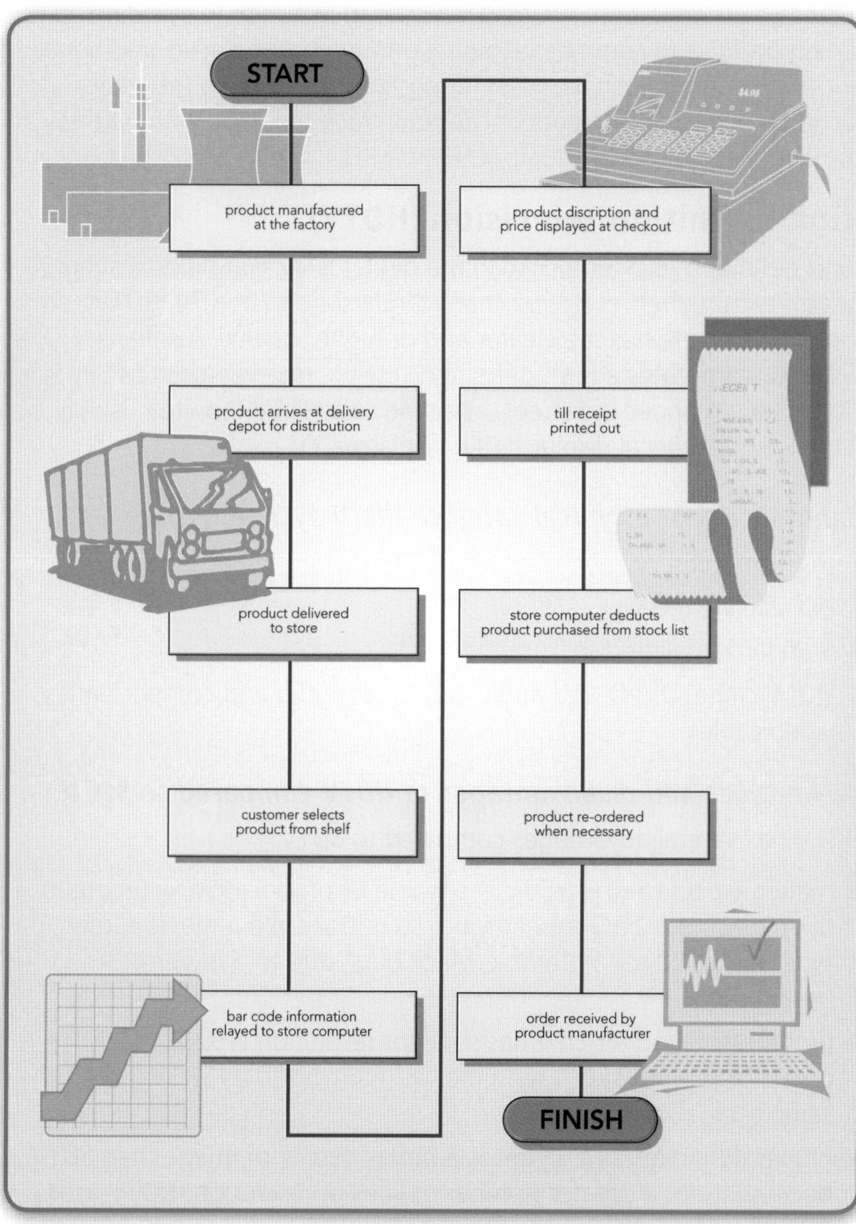

Figure 5.5: The electronic point of sale (EPOS) system

Stretch Activity

When EPOS is used with a store loyalty card, such as a Nectar card, a supermarket can build up a profile of your buying habits. This information can then be used to sell you products via email. Discuss the advantages and disadvantages of this system for the following:

- the supermarket
- you, the customer/consumer.

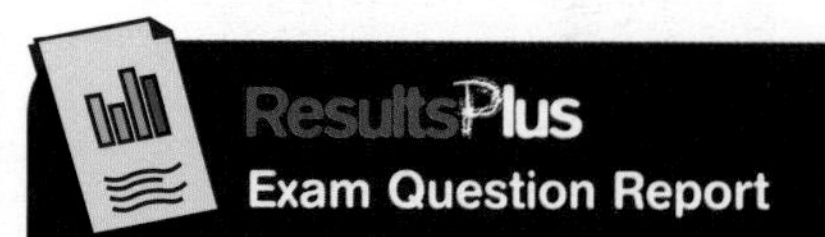

Describe how the barcode on a product allows the retailer to easily monitor sales of stock. (3 marks)

How students answered

Most students stated that a barcode is used to identify a product but did not then go on to describe the EPOS system. Many incorrect responses stated that the barcode contained the price.

52% 0-1 marks

Some students were able to give two stages in the EPOS system that allow the retailer to monitor stock levels.

22% 2 marks

Many students could give three stages in the EPOS system that allow the retailer to monitor stock levels.

26% 3 marks

Digital media and new technology

Objectives

- **Explain** the benefits of high-definition television (HDTV) and commercial digital printing.
- **Understand** the characteristics of the three types of commercial digital printing.
- **Describe** the process of transferring data using a Bluetooth® wireless personal area network.
- **Describe** the process of tracking products using radio frequency identification (RFID) tags.
- **Explain** the advantages and disadvantages of using radio frequency (RF) technology.

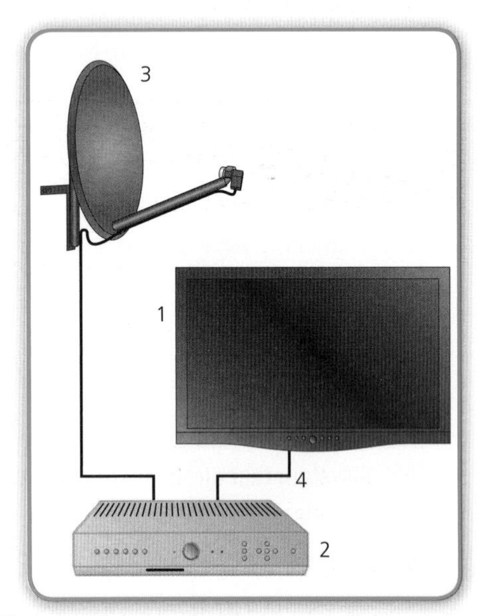

Figure 5.6: The components of a typical satellite HDTV system

We live in an age where technological developments are changing the way we do things. New digital media is fast replacing traditional methods. For example, digital television broadcasting systems such as high-definition TV are often chosen rather than standard definition TV, and commercial digital printing has replaced traditional printing methods such as offset lithography. Also, radio frequency technology means that people can now track products with wireless communications, using Bluetooth® and RFID tags.

High-definition television (HDTV)

High-definition television (HDTV) is a digital television broadcasting system with higher resolution than standard-definition TV (SDTV), so the pictures it creates are clearer and of higher quality. A satellite HDTV system incorporates a high-definition satellite receiver (such as Sky HD) or cables under the street, attached to a HDTV monitor, usually a slimline liquid crystal display (LCD) or plasma TV.

Components of a typical satellite HDTV system (Figure 5.6)

1 HDTV monitor
2 HD satellite receiver
3 Standard satellite dish
4 HDMI cable, DVI-D and audio cables, or audio and component video cables

Advantages and disadvantages of HDTV compared to SDTV

HDTV has several advantages compared to SDTV:

- higher aspect ratio – HDTV is capable of producing a widescreen format using a 16:9 aspect ratio, so it looks like a movie screen, whereas SDTV has letterbox 'black bars' due to a lower 4:3 aspect ratio.
- better resolution – HDTV has superior resolution (1920 × 1080 pixels) displaying about ten times as many pixels as SDTV (about the same as analogue TV).
- higher definition – HDTV gives a better quality of image than SDTV because it has a greater number of lines of resolution (HDTV uses 1080 lines, over twice as many lines as SDTV).
- higher frame rate – this means how many times a TV creates a complete picture on the screen every second. SDTV frame rates range from 24 to 60 frames per second. HDTV displays 60 frames per second, even though it has a much higher resolution that SDTV. The higher the frame rate, the better the quality of the picture you see.

At present, the main disadvantage of HDTV is cost, as to use it you will have to bear the additional cost of buying an HD-ready television and the higher subscription cost of subscribing to an HD service.

Figure 5.7: HDTV 'brings images to life' due to its superior resolution

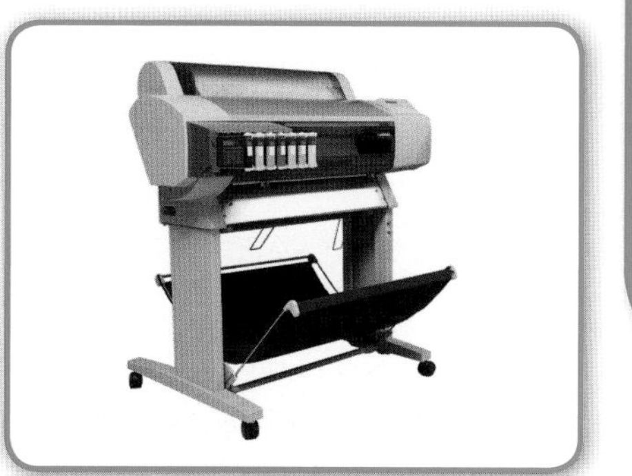

Figure 5.8: Large posters can be produced quickly using large-format digital printing

Commercial digital printing

Commercial digital printing uses electronic files in print production rather than the traditional printing plates used in offset lithography. Because there is less initial set-up, digital printing is useful for rapid prototyping (see page 42) and is cost-effective for small print runs.

Digital printing also means that you can customise mass-produced documents using digital print technology. Instead of producing 10,000 copies of a single document, delivering a single message to 10,000 customers, digital printing could print 10,000 unique documents with customised messages for each customer.

There are three main types of commercial digital printing.

- **Print-on-Demand (POD)** allows for small amounts of printing to be done, which is perfect for companies that need to constantly update their business documents, such as brochures. Large-format digital printing can produce large banners and posters for promotions. POD can also be used to make short print runs of books.
- **Variable Data Printing (VDP)** is a customised and personalised type of digital printing. Databases containing specific consumer information make each piece of the same mail design personalised with their name and address. Consumers appreciate personalised mail as it is less general.
- **Web-to-Print** digital printing allows for direct mail pieces to be customised and personalised on-line. Customers can choose images and photographs and add personalised text to items such as greetings cards. A proof is shown online and when the piece is ready, one click sends it to the printer for an extremely quick turnaround.

Commercial digital printing differs from offset lithography, flexography and gravure printing in several ways. With commercial digital printing:

- every print can be different because printing plates are not required, as in traditional methods
- there is less wasted paper and printing ink because there is no need to bring the printed image 'up to colour' and check for registration and position
- the ink or toner does not soak into the paper as with normal printing inks, but forms a thin layer on the surface, so it dries quickly.

However, whenever a very high-quality printed product is required, traditional printing processes are still used.

Support Activity

1. Why would digital printing be good for producing 500 copies of a flyer to promote a local band?
2. Go to www.moonpig.com and try producing a personalised greetings card.

Stretch Activity

Visit a local electrical retailer and see the difference between HDTV and SDTV. Is the cost of these systems worth it? What will it take for HDTV systems to become more affordable for most people?

ResultsPlus
Build Better Answers

Explain **two** advantages of high-definition television (HDTV) compared with standard-definition television (SDTV). (4 marks)

▲ **Basic answers (0-1 marks)**
Give only one generic advantage of HDTV such as 'better picture', with no justification of how it produces a better picture.

● **Good answers (2-3 marks)**
Give two advantages of HDTV that make it better than SDTV, but some responses are not fully justified.

■ **Excellent answers (4 marks)**
Achieve full marks when two advantages of HDTV are fully justified.

Bluetooth®

Bluetooth® is a wireless system for connecting several devices together and exchanging data over short distances using short-range radio frequency (RF) waves. This creates a network known as a personal area network (PAN) or piconet. Bluetooth® doesn't require line of sight between communicating devices: for example, walls won't stop a Bluetooth® signal. You can connect up to eight devices such as mobile phones, laptops, personal computers, printers, Global Positioning System (GPS) receivers, digital cameras, and video game consoles at the same time within a 10-metre radius, such as your house.

The Bluetooth® frequency band has been set aside by the international agreement for the use of industrial, scientific and medical devices (ISM), so Bluetooth® does not interfere with other communication systems, such as television.

However, there may be other devices in your house that use this same frequency band, such as baby monitors and garage-door openers. To deal with this, Bluetooth® uses a radio technology called frequency-hopping spread spectrum, which chops up the data being sent and transmits chunks of it on up to 79 frequencies. This makes it rare for more than one device to be transmitting on the same frequency at the same time and avoids interference. Bluetooth® transmitters automatically change frequencies 1,600 times every second, so more devices can make full use of a limited slice of the radio spectrum.

When Bluetooth® devices come within range of one another, an automatic electronic conversation takes place to determine whether they have data to share or whether one needs to control the other. This forms a piconet. Once this piconet is established, the members randomly hop frequencies in unison so they stay in touch with one another, while avoiding other piconets that may be operating in the house.

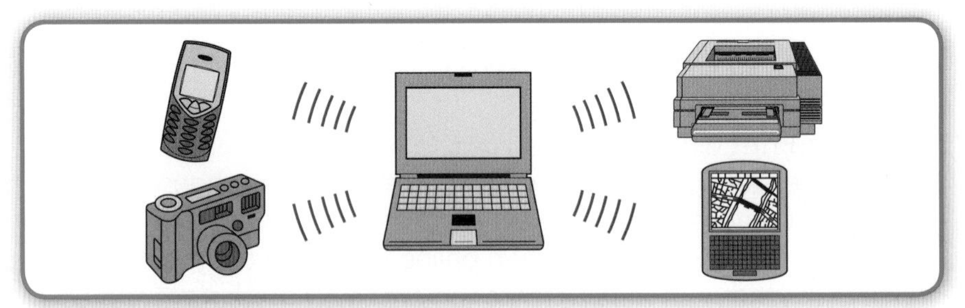

Figure 5.9: A Bluetooth® network

One disadvantage: security

As in any wireless network, security is a concern as hackers can intercept signals and spy on or get remote access to a Bluetooth® device. Two particular security issues with Bluetooth® devices are 'Bluejacking' and 'Bluebugging'.

- Bluejacking involves Bluetooth® users sending a business card (in the form of a text message) to other Bluetooth® users within a 10-metre radius. If the user doesn't realise what the message is, they might allow the contact to be added to their address book. The contact can then send messages that might be automatically opened because they're coming from a known contact.

- Bluebugging is more of a problem, because it allows hackers to remotely access a user's Bluetooth®-enabled phone and use its features. This could include placing calls and sending text messages, without the user realising what is going on.

To beat these problems, Bluetooth® has a range of security systems, including special authorisation and identification procedures. These require the user to make a conscious decision to open a file or accept a data transfer, without it simply happening automatically.

Radio frequency identification (RFID)

Radio frequency identification (RFID) is a method of identifying products by sticking tags to them; the tags store data that can be retrieved by an electronic reader. This method is often used to track pallets of products from the manufacturer to the retailer. It also has applications in libraries, where a tag is used on library books to identify the book and the person borrowing it. This proves useful as a security system to stop books being stolen.

RFID technology is based on the transmission and reception of radio frequency (RF) signals between a transmitter (the reader) and a transponder (the tag), which is attached to the product. In most cases the transmission is two-way: the transmitter sends signals, which the transponder receives. The transponder then transmits a response signal that is received by the transmitter. The information from the transmitter can then be used to identify the transponder and any product it is attached to.

RFID readers and tags work on many frequencies and are available in either two formats: active or passive.

- **Active transponders** are battery-powered and can be read over a greater distance, but are usually expensive. They are generally used on items such as vehicles, to make automatic road toll payments, and on shipping containers. Being battery-powered, they have a limited lifetime.
- **Passive transponders** take the power they need to respond from the electromagnetic signals transmitted by the reader, but the strength of these signals is quite weak, giving them a limited range. They are commonly used where the transponder needs to be thin and can be easily made into a label: for example, to track pallets or for security systems in libraries.

The main concern with RFID tags is that they could be used for the wrong reasons. If RFID tags are attached to all product, these products could easily be tracked to your own home. Some people argue that this is an invasion of privacy, and are concerned how manufacturers and marketing companies could use this information.

ResultsPlus
Build Better Answers

Explain **two** disadvantages of using a wireless Bluetooth® system when communicating business information. (4 marks)

Basic answers (0-1 marks)
Give only one generic disadvantage of Bluetooth® such as 'security issues', with no further explanation.

Good answers (2-3 marks)
Give two disadvantages of using Bluetooth® for communicating business information, but some responses are not fully justified.

Excellent answers (4 marks)
Achieve full marks when two disadvantages of Bluetooth® are fully justified.

Figure 5.10: A radio frequency identification (RFID) tag and reader

Support Activity

From 2006, all new UK passports now contain an RFID tag. These 'e-passports' are used at airports to identify you as you arrive in and depart from a country. Do you think that this type of system should be used to track people? Is privacy most important, or does security take priority?

Stretch Activity

You will often see people walking down the street or driving in cars using Bluetooth® headsets. Use the internet to find out how these devices work.

CAD/CAM technology

Objectives

- **Explain** the benefits of using computer-aided design (CAD) when designing and developing graphic products, including 2D drafting, desktop publishing (DTP) and 3D virtual modelling and testing.
- **Explain** the benefits of using computer-aided manufacture (CAM) when manufacturing graphic products, including laser cutting and engraving and vinyl cutting.
- **Understand** the impact of CAD/CAM on the workforce.

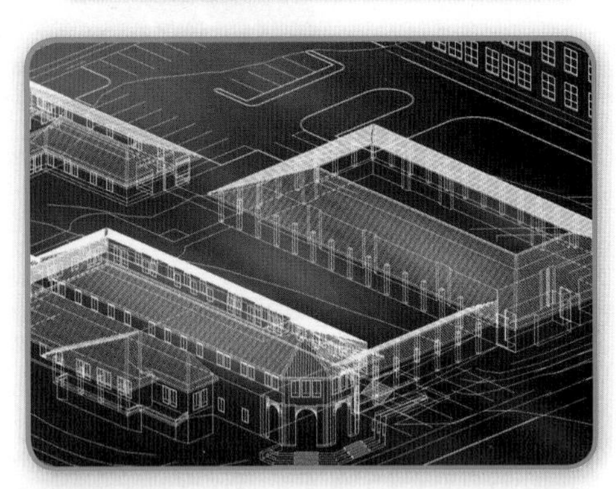

Figure 5.11: Architects make great use of CAD drawings when planning their buildings

Apply it!

You will be able to use CAD in your Unit 1: Creative Design and Make Activity.

You can use 2D drafting on ProDesktop or 2D Design to produce your final design proposal in the design activity. You can also use ProDesktop to produce 3D models of your designs to aid the development of the product. You can use a DTP package to develop designs for many printed products.

Computer-aided design and computer-aided manufacture technologies have revolutionised the design, development and manufacture of products. CAD reduces the time it takes to develop a new product, while CAM enables products to be made efficiently and quickly. However, both technologies have had an impact on the modern workforce.

Designing graphic products using CAD

The use of computer-aided design in the design and development of graphic products is now the industry standard, replacing the need for laborious and time-consuming manual drawing and cut-and-paste techniques. The possibilities created by this new technology have led to rapid developments in the printed media, the internet, animation and special effects for TV and film.

2D drafting

2D drafting using CAD is an important industrial process used in many graphic products, including packaging, product, automotive and architectural design. 2D drafting is used to produce detailed engineering drawings of components and layouts of buildings.

CAD has become a major factor in lowering the cost of developing a product, and greatly shortening the design process and the product's time to market.

Advantages and disadvantages of CAD for developing designs

Here are some of the advantages of using CAD.

- Designs can be modified quickly on screen without the need to re-draw the entire product or component.
- The designer has access to a large library of standard components which do not have to be drawn from scratch, but can be modified to suit specific uses.
- Designs can be saved electronically, easily retrieved and sent via email to clients for approval.
- Design data can be directly outputted to computer-aided manufacturing (CAM) equipment such as rapid prototyping (RPT) machines, which produce 3D models for testing.

The main drawbacks to CAD are the massive learning curves that users have to undergo to understand how to use a piece of software. CAD packages are often complicated to begin with – you may remember how you felt when learning how to use a simple wordprocessing package on your computer at home or school. However with the right training and experience, operators soon become skilled at using them. At the same time, user interfaces have become easier to use in recent years and a wide range of training courses is available for the industry standard software packages (e.g. AutoCAD and ArchiCAD).

Desktop publishing

Desktop publishing (DTP) combines the features of wordprocessing, graphic design and printing in one package. All modern print media use DTP to lay out pages electronically, rather than using the traditional process of typesetting. This textbook, for example, was designed using DTP and then sent to the printers for commercial printing. The designer could be located in one place while the printers could be based elsewhere in the UK, or even overseas; as everything is saved onto a computer, data can be sent anywhere.

Advantages of DTP for developing designs

With DTP, you can:

- set up documents easily, including customised page measurements and number of pages
- add page layout grids and guides on screen to help you position text, graphics and images (these are invisible when printed)
- add colour separations and 'bleed' areas, to aid the commercial printing process
- manipulate text, graphics and images easily
- choose from a wide range of typefaces to suit different graphic styles, and graphic tools to produce customised graphics
- zoom in and out for attention to detail, and preview pages before printing
- save documents electronically in various formats and send them electronically to the printers for commercial printing.

3D virtual modelling and testing

A three-dimensional image can give a more realistic impression than a two-dimensional image. Many designers will construct new products on screen which, with skill, can easily be modified and manipulated. Product design teams can use 3D modelling to design and develop a new product much faster, saving developmental costs and time to market. Often the virtual products can be tested and evaluated without actually being manufactured, and design data can be directly outputted to a CAM system for modelling prototypes.

The rapid development in computer gaming has produced an impressive range of computer-generated characters, which can interact with lifelike scenarios and landscapes. Most feature films released nowadays use computer-generated special effects, and even actors. In 3D animation, first a wireframe model is created on screen, which can be viewed in all directions thanks to camera angles built into the computer software. The wireframe model is then rendered, so that it has an appropriate surface finish or texture. This involves wrapping a 'skin' around the wire frame to give a photo-realistic image. Architects often use photo-realistic virtual models to show clients what their new building will look like; animated 'walk-throughs' of interior designs can give the client a real feel for the finished product.

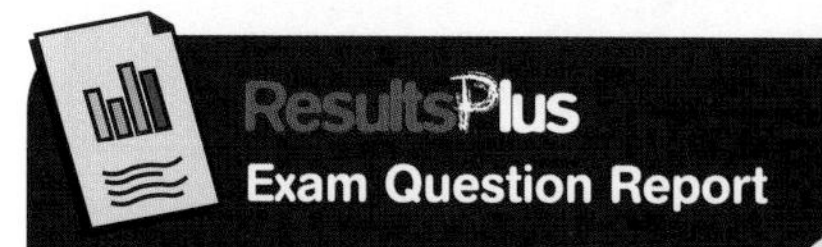

Describe why computer-aided design (CAD) was used to design this product instead of a drawing board and technical drawing equipment. (3 marks)

How students answered

Most students only stated one feature of a CAD system but failed to provide reasons why this feature was used in preference to traditional methods.

54% 0-1 marks

Many students were able to state two benefits of using CAD such as 'easily modified on screen' and 'saved electronically', or fully justified one benefit.

54% 2 marks

Some students could identify at least two features of a CAD system and justify why it was used instead of a drawing board. Alternatively, students achieved full marks by identifying one feature with a detailed justification with two clear marking points.

13% 3 marks

Figure 5.12: Architects made extensive use of virtual modelling when planning the London 2012 Olympic park

Figure 5.13: The role of the designer has altered dramatically with the use of CAD

Making graphic products using CAM

Since the Industrial Revolution, the manufacturing process has undergone many dramatic changes. One of the most dramatic is the introduction of computer-aided manufacture, a system of using computer technology to help the manufacturing process. CAM allows data from CAD software to be converted directly into a set of manufacturing instructions.

With a CAM system controlling the production process, a factory can become highly automated and create products with a level of precision that would not be possible if everything were done by hand.

Laser cutting and engraving

Laser cutting involves using a high-powered laser to cut materials, often with intricate designs. The laser is directly controlled by computer using a CAD/CAM system, so there is no need for expensive tools. The laser melts, burns or vaporises away the unwanted parts of the material, leaving an edge with a high-quality surface finish. This process is ideal for manufacturing signage, for example, from sheet acrylic. By altering the depth of cut of the laser beam, you can also engrave materials.

Advantages and disadvantages of laser cutting over mechanical cutting

The advantages of laser cutting include:

- lack of physical contact with the material produces a clean edge
- precision, as there is no wear on the laser and it is computer-controlled
- reduced chance of warping the material as only a small area is affected by the heat
- some materials or features are difficult or impossible to cut using traditional methods.

Disadvantages include:

- very expensive machinery
- high energy consumption.

Vinyl cutting

Vinyl is the common name for plasticised PVC. Vinyl stickers or graphics are ideal for one-off or batch production; they might be used for anything from an individual sign for a shop or restaurant frontage, to a series of movie adverts covering the backs of buses.

Commercial sign-makers usually have banks of images and accurate dimensions so that they can batch-produce vinyl graphics such as logos or trademarks for large companies. However, many schools and colleges have access to CAD/CAM facilities for cutting vinyl that directly replicate the commercial process.

Apply it!

You will be able to use CAM in your Unit 1: Creative Design and Make Activity.

If you have access to a laser cutter or engraver, you can use it to produce up to 50% of the components in your product. You can use vinyl cutting to enhance your final model by, for example, adding a brand name and logo to a 3D model.

Figure 5.14: Vinyl graphics proved successful when marketing and personalising the new Mini

How the process works

- The image is designed on computer and sent in digital form to the vinyl cutter for contour cutting.
- Once the cutting is complete, the background vinyl is removed using a process called 'weeding', leaving only the required graphic.
- The adhesive is already on the rolls of vinyl. Contact or impact adhesive has been lightly coated on the back and covered with a treated backing paper to protect it. This makes the vinyl easy to peel off.
- Finally, a layer of application tape is applied. The application tape will adhere to the vinyl so that it can be removed from the backing sheet and put onto the required surface. Vinyl graphics have a soft, flexible quality that that means they will stick to a variety of contours. Once the graphic has been applied, the application tape can be removed leaving only the vinyl graphic in place.

The impact of CAD/CAM on the workforce

The modern workforce has had to become skilled, trained and, above all, flexible.

A significant number of people working in manufacturing are unskilled and low-paid machine operatives, assemblers or packers. As automated machinery develops, and the use of robots widens, there will be even less human involvement in the manufacturing and production processes of the future. These changes can create emotional and other problems at work.

Effects of CAD

Designing used to be carried out in large drawing offices, with lots of people working on the different components of the same job. Skilled technicians and draughtspeople hand-drew all of the technical information required to manufacture the product.

Now, a small team of highly skilled computer designers can perform the same tasks more efficiently and in less time. Many draughtspeople have had to retrain to become computer literate and familiar with industry standard software packages. Those unwilling or unable to retrain either have fewer work opportunities or become redundant.

Effects of CAM

Traditionally, workers in manufacturing industries have been highly skilled machine operators. School-leavers followed apprenticeship courses in the employer's factory to train in the operation of a particular piece of machinery. Young people would often serve their apprenticeship, then work for the same employer until they retired.

Operators of CAM machinery now have to be trained to operate different machines and carry out many processing tasks within their manufacturing cell or work centre. Gone are the days of narrow specialisation when, for instance, lathe operatives were not allowed to set up and use another person's machine in the factory. People now have to work as part of a team, offering help to others if required.

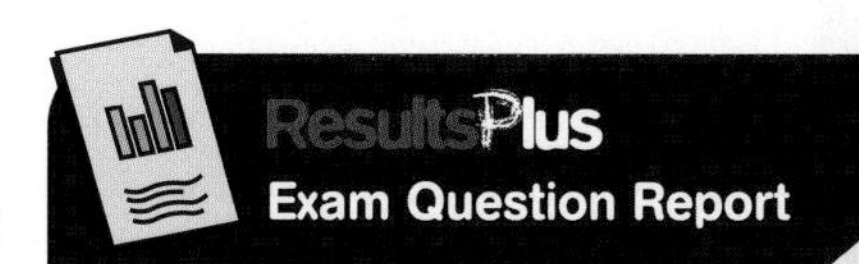

Explain two reasons why CAD/CAM can have disadvantages for the employees of companies that produce graphic products. (4 marks)

How students answered

Most students could only give one general disadvantage such as 'people lose their jobs' without any explanation as to why CAD/CAM may cause a reduction in the size of the workforce.

62% 0-1 marks

Many students were able to include two valid disadvantages, such as specialist CAD teams and automated CAM machinery. However, only one reason was justified in sufficient detail.
Remember that, in an 'explain' question, you have to make a valid point and then go on to justify it. You cannot achieve full marks without any justification.

28% 2-3 marks

Some students correctly identified two appropriate disadvantages, with each being fully justified.

10% 4-5 marks

Support Activity

If you have access to CAM machinery, learn how to use it by designing a small product in a CAD package, setting up the machine with help from your teacher or technician and making the product. Now try to make the same product by hand. Which one took the least amount of time to make? Which one looks the best?

Know Zone
Chapter 5 Technology

The modern world relies on technology. Computers are used throughout the world to communicate, design and help manufacturing systems become more efficient. Technology plays an important role in all of our lives from our mobile phones to the high definition TVs that entertain us. Can you think of a world without technology? It's hard to imagine but technology has had a dramatic impact upon the modern workforce.

You should know...

the following about technology:

Technology	Characteristics	Processes	Advantages and disadvantages	Uses
Email	✓	✓	✓	✓
EPOS	✓	✓	✓	✓
Internet	✓	✓	✓	✓
Bluetooth	✓	✓	✓	✓
HDTV	✓	✓	✓	✓
Digital printing	✓	✓	✓	✓
RFID	✓	✓	✓	✓
CAD 2D drafting	✓	✓	✓	✓
DTP	✓	✓	✓	✓
3D virtual modelling	✓	✓	✓	✓
CAM Laser cutting/engraving	✓	✓	✓	✓
CAM Vinyl cutting	✓	✓	✓	✓

Key terms

Questions relating to technology will require you to know technological processes, why they are used but also the impact they have had upon the modern workforce.

Property	Meaning
Characteristics	Recognisable features that help to identify or differentiate one process from another. For example, why would commercial printing be used instead of offset lithography for printing a company's mail outs?
Processes	A description of the series of actions needed to produce a product or component. For example, knowing the stages in the EPOS system.
Advantages and disadvantages	Qualities and features favourable to success or failure. For example, what are the advantages of using laser cutting rather than traditional cutting processes?

Grade E-C range question: Explain **one** disadvantage of using e-mail for communication. (2 marks)

Student answer ■	Examiner comments	Build a better answer △
E-mail is a great way of communicating over large distances very quickly. (0 marks)	The question clearly asks for one disadvantage so no marks can be awarded.	Email is impersonal (1 mark) so messages can be misinterpreted as you are not communicating face to face. (1 mark)

Overall comment: Please read each question carefully before you respond. Some questions will ask you for both advantages and disadvantages and you cannot achieve full marks if you don't mention both.

Grade C-A range question: Describe the process of making vinyl stickers for a new shop sign. (4 marks)

Student answer ●	Examiner comments	Build a better answer △
The stickers are made using a CAM machine (1 mark). *The stickers are then removed from the backing sheet and stuck to the sign.* (1 mark)	A basic response that is lacking in any real detail needed to achieve full marks. How are the stickers made using CAD/CAM ? What is the process of removing and applying the stickers?	The design is created in a CAD programme (1 mark) then outputted to a CAM vinyl cutter for an accurate cut (1 mark). The unwanted vinyl is removed (1 mark) and application tape is placed over the letters needed. (1 mark) The tape removes the letters from the backing sheet so that they can be applied to the sign in one go. (1 mark) The stickers can then be smoothed out to remove any air bubbles. (1 mark)

Overall comment: Notice that on the 'build a better response' there are actually six rewarded points being made. Sometimes, it is good practice to include more detail than is needed.

'Stretch and Challenge' A/ A* question: Evaluate the use of CAD/CAM systems in modern manufacturing. (6 marks)

Student answer ●	Examiner comments	Build a better answer △
CAM has made manufacturing more efficient because machines are controlled by computers (1 mark) *which also makes all operations more accurate than humans.* (1 mark) *Using CAD you can design a product without actually having to model it which saves time and money* (1 mark). *This means that a product can be put into shops quicker than having to test lots of hand-made prototypes.* (1 mark)	A good response but only four relevant points have been made. All the points refer to advantages of CAD/CAM which does not make an 'evaluation' because there are no disadvantages. Take care when using phrases such as 'machines are more accurate than humans' as a skilled craftperson is very accurate. However, a machine could make the same product faster.	CAD/CAM lets products reach the market quicker (1 mark) by reducing development time and saves money. (1 mark) CAD systems lets the designer build and test 3D ideas on screen (1 mark) without actually making costly 3D models. (1 mark) CAM systems are extremely accurate and efficient (1 mark) but can reduce the workforce to 'machine-minders' instead of skilled craftspeople. (1 mark)

Overall comment: An 'evaluation' response has to address both advantages and disadvantages to achieve full marks. This 'build a better answer' outlines five advantages and just one disadvantage – but both are present!

Chapter 6 Sustainability
Minimising waste production

Objectives

- **Describe** how waste is minimised throughout the product life cycle using the '4 Rs'.
- **Understand** the benefits to society and the environment by using the '4 Rs'.

Sustainable product design

Sustainability means safeguarding the world, for ourselves and for future generations. Sustainable product design involves our need to design and manufacture products using energy and materials in a way that keeps our use of finite resources, waste production and pollution as low as possible. Perhaps the most important economic factor for a designer of sustainable products to consider is that waste is lost profit.

What is the product life cycle?

A product's life cycle refers to the key stages in the life of a product 'from the cradle to the grave', so from the extraction of the raw materials it is made from right through to how it is disposed of at the end of its life.

The '4 Rs'

There are four simple options to consider when deciding how to minimise waste production throughout the product life cycle, referred to as the '4 Rs':

- **reduce** materials and energy
- **reuse** materials and products where applicable
- **recover** energy from waste
- **recycle** materials and products, or use recycled materials.

Life cycle stage	Environmental points
Raw materials	• Use less material • Use materials with less environmental impact • Consider recyclable materials • Adhere to relevant laws
Manufacture	• Reduce energy use • Simplify processes where appropriate • Reduce waste • Use natural resources efficiently
Distribution	• Reduce or lighten packaging • Reduce mileage of transportation to customer
Use	• Increase durability of product • Encourage refill consumables where appropriate • Use 'green' credentials as a positive marketing strategy • Promote use of efficient use of product
End-of-life	• Make reuse and recycling easier • Reduce waste to landfill

Table 6.1: Key environmental considerations for the product life cycle

Reduce

For all designers, one of the first priorities for sustainability should be to reduce the quantities of any material chosen, wherever possible. Packaging designers must make the most of the materials they use to package a product, in order to minimise the use of resources. By reducing, they will also save on costs and improve their profits.

Manufacturers are asked to reduce packaging use under the UK Producer Responsibility Obligations (Packaging Waste) Regulations 1997. The Government's Envirowise programme suggests that manufacturers should:

- consider the materials and designs they use
- examine ways of eliminating or reducing the packaging requirement of a product – changes in product design, improved cleanliness, better handling, just-in-time delivery, bulk delivery, etc.
- make the best use of packaging, by matching packaging to the level of protection needed.

Reductions in the packaging size of detergents and fabric conditioners offer one good example. Technological developments have led to more concentrated forms of these products, resulting in less packaging being needed per measure of detergent.

Reuse

Reusing products reduces the extraction and processing of raw materials to produce new items, and reduces the energy and resources required for recycling. A number of companies use returnable or refillable containers for products: for example, companies that deliver milk in glass bottles.

Refillable containers can offer environmental benefits. However, more resources may be needed to manufacture and distribute them, so that they can stand up to repeated use. Using extra resources like this may be balanced out by reuse, but only in local distribution and collection schemes. If reuse is to be worthwhile, the cost of collecting, washing and refilling must be less than the cost of producing a new container.

Recover

The manufacture of any product requires the use of energy. If the product is simply discarded and buried (landfill), all of this energy is lost. Waste that cannot easily be recycled but can burn cleanly can be incinerated in specialised power stations, generating electricity and providing hot water for the local area. Recovery is not as effective as reduction, but adopting such technology does mean that less finite fossil fuel is needed to generate electricity in conventional power stations. In Sweden 47% of waste is recovered in energy from waste plants; in the Netherlands, 34% is recovered. The Tetra Pak™ education service has claimed that you could run a 40 W bulb for an hour and a half on the energy released when one of their cartons is burnt.

Recycle

Recycling takes waste materials and products and reprocesses them to manufacture something new and useful. Some materials, such as paper and board (see Chapter 1), can be made into the same products again, such as newsprint into newspapers; other materials can be made into something completely different, such as plastic vending cups into pencils.

Recycling is important in a modern consumer society, where millions of tonnes of waste are disposed of in landfill sites or incinerated. However, recycling is only part of the answer, as the product has already been made. Using less material and energy in the first place will significantly reduce a product's environmental impact.

Designers and manufacturers are becoming increasingly aware of the possibilities of using recycled materials, especially where consumers do not mind a slight degradation in the quality of the product. For example, using recycled paper in a notepad may result in a 'greyer' finish, but consumers may accept this because they are more aware of environmental issues and want to 'do their bit'.

However, it is now extremely important for a designer to consider design *for* recycling. Here the designer must consider the product life cycle from the outset, to fully appreciate how it will be disposed of when it has come to the end of its useful life. If designers create components that are easy to take apart and use materials that can be recycled after use, they can contribute to minimising waste production.

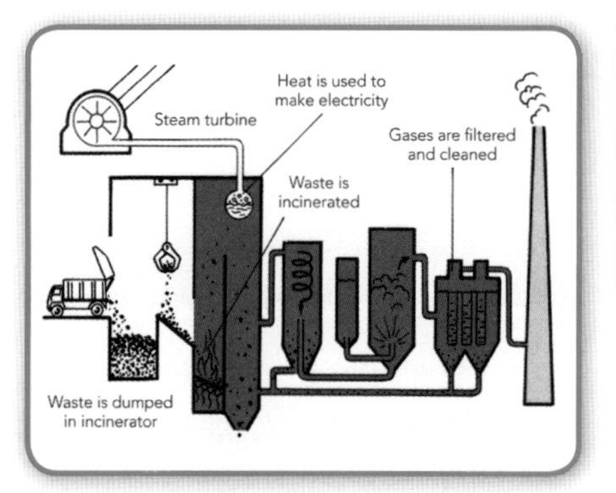

Figure 6.1: Recovering energy from waste

Support Activity

1. Go back to Chapter 1: Materials and components, and cross-reference your notes with this section, looking at the environmental issues relating to specific materials.
2. Study a piece of packaging that you are familiar with. How could the materials be reduced to minimise waste production?
3. Would you pay more for a product that was 'environmentally friendly'? Explain your answer.

Stretch Activity

Conduct a study of the waste generated by your school or college in a typical day. Where does the majority of it come from (e.g. litter, photocopying, canteen)? How could this waste be minimised?

Renewable sources of energy

Objectives

- **Explain** the advantages and disadvantages of using wind and solar energy.
- **Understand** the process of converting wind energy into electrical energy.
- **Understand** the process of converting solar energy into either heat or electrical energy.
- **Explain** the advantages and disadvantages of using biomass and biofuels.
- **Understand** the importance of developing biofuels for transportation.

What are renewable sources of energy?

Renewable energy is generated from natural resources such as sunlight, wind, rain, tides, and geothermal heat, which are naturally replenished. Non-renewable sources of energy come from coal, oil and gas, which are finite resources that cannot be replenished.

Wind energy

A wind turbine is used to convert the kinetic energy of the wind into mechanical energy, which is in turn converted into electricity.

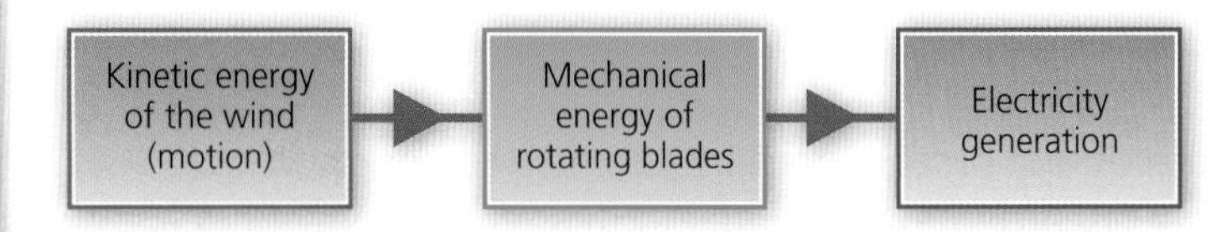

Groups of large turbines, called wind farms, are the most cost-efficient way to capture energy from the wind. In the UK, wind farms are usually situated in mountainous, rural areas with high wind speeds, or in coastal waters to capture offshore winds. The turbines used in wind farms for commercial production of electricity are usually three-bladed horizontal-axis wind turbines (HAWT). These can be pointed into the wind by computer-controlled motors, and are equipped with shut-down features to avoid damage at high wind speeds.

Electricity production using a wind turbine

1 The energy in the wind is converted to rotational motion by the rotor.
2 When the blades turn, the rotor turns a shaft, which transfers the motion into the nacelle (see Figure 6.3).
3 The slowly rotating shaft enters a gearbox that greatly increases the rotational shaft speed.
4 The output shaft is connected to a generator that converts the rotational movement into electricity at medium voltage (hundreds of volts).
5 The electricity flows down heavy electric cables inside the tower to a transformer, which increases the voltage of the electric power to the distribution voltage (thousands of volts).
6 The distribution voltage flows through underground lines to a collection point, where power is combined with that from other turbines.
7 The electricity is sent to a substation, where the voltage is increased to transmission-voltage power (hundreds of thousands of volts) and sent to the National Grid for use throughout the UK.

Figure 6.2: Wind farms including many turbines to commercially produce electricity

Advantages	Disadvantages
• flexibility – can be used in large-scale wind farms for National Grid as well as in small individual turbines for providing electricity to rural houses or isolated locations • non-polluting, environmentally friendly and sustainable – produces more than 50 times as much energy over its lifetime as is consumed by its construction and installation • produces low-cost power if developed commercially, involving low running and maintenance costs • can be installed off shore to minimise visual impact and take advantage of the constant breezes	• can only provide a small amount of total energy needs due to the small amount of turbines available • relies on winds • unsightly on-shore wind turbines and wind farms spoil picturesque landscapes • extremely expensive to build off-shore wind farms • infrastructure required for wind farms causes some damage to landscape, e.g. access roads for maintenance • controversial – noise and vibration of moving turbine has potential to affect local community and livestock • affects environmentally sensitive coastal sites, e.g. those with lots of nesting birds

Table 6.2: Advantages and disadvantages of using wind energy

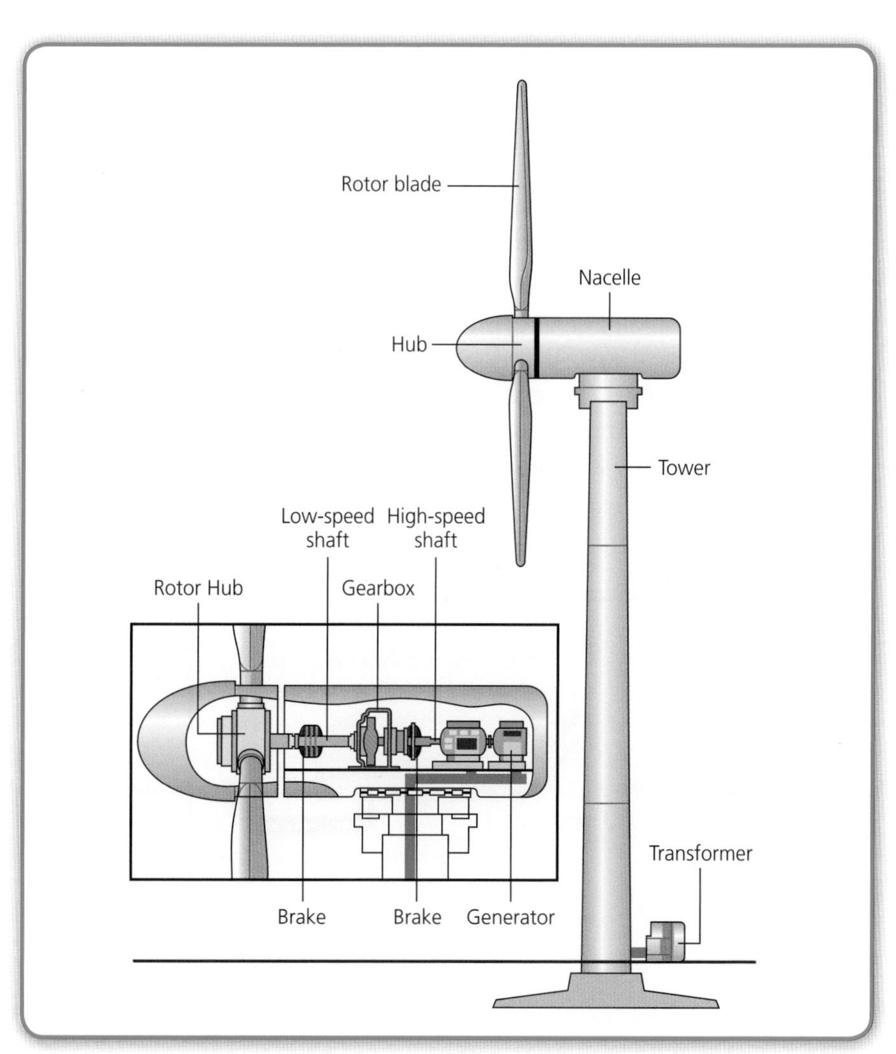

Figure 6.3: A horizontal-axis wind turbine (HAWT)

Figure 6.4: Satellites use solar panels to catch the sun's rays to provide power to the equipment on board

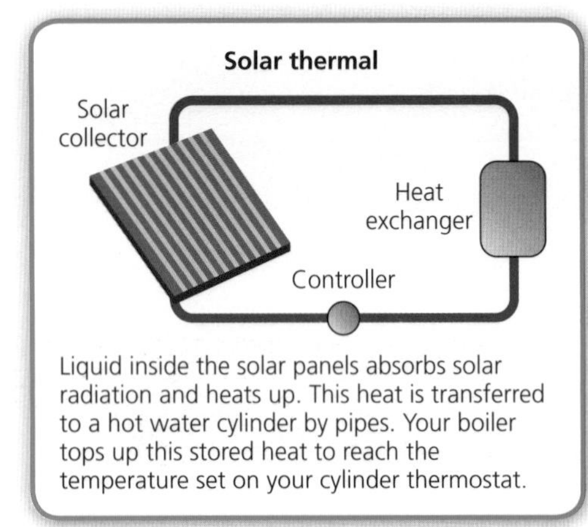

Figure 6.5: Solar panel hot water system

Solar energy

With solar energy, the energy of sunlight is converted directly into electricity using a device called a photovoltaic cell. Many cells can be assembled together to form a solar panel.

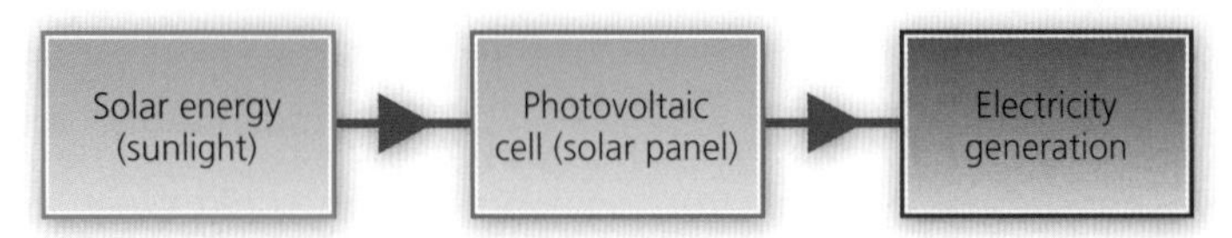

Solar panels are a useful way of providing electricity to remote areas where it may not be viable to lay high voltage cable. The best example of the importance of solar energy to provide electricity in remote locations is on satellites orbiting the Earth.

It is quite possible for a household to run completely off electricity from the use of solar panels, yet this is unlikely to happen in reality: for the average homeowner, the costs involved in setting this up would be too high. However, in the average home, solar electricity can provide a substantial amount of electricity, reducing future energy bills.

Electricity production using a solar panel

The way that solar panels produce electricity involves some complex science, but here is a brief summary.

1 Atomic particles called photons in sunlight hit the solar panel and are absorbed by materials known as 'semiconductors', such as silicon – the material used in microchips
2 Other particles called electrons (negatively charged) are knocked loose from their atoms, and flow through the material. Because of the special nature of the solar cells, the electrons can only flow in one direction, and this produces electricity.
3 The solar panel converts the solar energy into a usable amount of direct current (DC) electricity, which is the sort of electricity you can use to charge batteries.

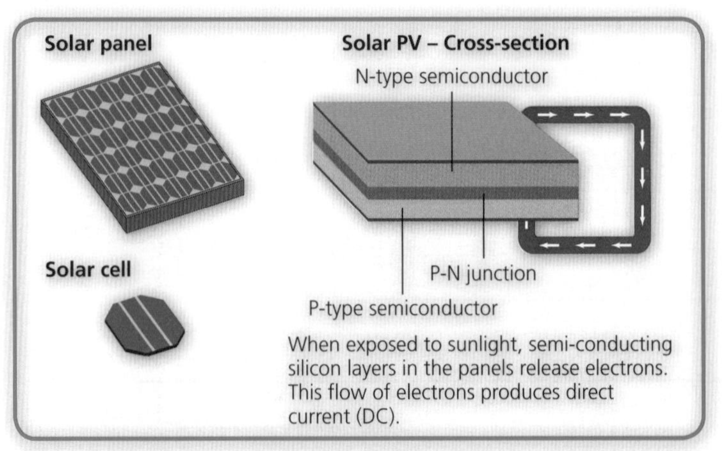

Figure 6.6: Cross-section through a solar panel

Using solar panels to heat water is becoming increasingly popular due to the savings in energy and costs that this brings. A good solar hot water panel system can provide an average household with around a third of its annual hot water supply. To maximise efficiency, a solar panel hot water heating system and a solar electricity system can be combined.

Advantages	Disadvantages
• huge amounts of energy available from the sun • pollution-free during use • low operating costs and very little maintenance required after initial set-up • economically competitive, especially for isolated or remote regions • produces enough electricity for the National Grid to cope with peak demand times • local grid-connected solar electricity systems can be self-sufficient	• relatively expensive set-up costs • not available at night and may be unavailable due to poor weather conditions, so a storage power system is needed • energy lost by converting DC current generated into AC current needed for use in the National Grid

Table 6.3: Advantages and disadvantages of using solar energy

Solar electricity & solar water heating
It's important that the panels are mounted on south-facing facades (or anywhere from East to West through South.) This lets panels absorb maximum daylight hours. Solar electricity can either be connected to the National Grid, or connected to a battery.

Figure 6.7: Combined domestic solar electricity and solar water heating system

Support Activity

1 Why is it a good idea to build large-scale wind farms in coastal waters around the UK?

2 Why are solar panels on the roofs of houses more common in European countries like Spain than in the UK?

Stretch Activity

Research the advantages and disadvantages of using fossil fuels and nuclear energy in relation to using renewable sources of energy.

ResultsPlus
Build Better Answers

Describe the process of generating electricity using a wind turbine. (4 marks)

▲ Basic answers (0-1 marks)
Give a generic response such as 'blades rotate in the wind, which in turn generates electricity'.

● Good answers (2-3 marks)
Describe up to three specific stages in the process of generating electricity using a wind turbine.

■ Excellent answers (4 marks)
Achieve full marks when four specific stages are described with some technical detail.

Biomass and biofuels

Support Activity

1 Why are algae a good base material for producing biofuels?

2 Do you think the ecological damage of growing crops for biofuels outweighs the environmental benefits of using biofuels? Explain your answer.

Stretch Activity

Research the advantages and disadvantages of fossil fuels like petrol and diesel for transportation. How do they compare with biofuels?

ResultsPlus Build Better Answers

Evaluate the use of biomass for producing biofuels for transportation. (6 marks)

Basic answers (0-2 marks)
Give up to two advantages of using biofuels rather than fossil fuels to run vehicles.

Good answers (3-4 marks)
Explain two advantages to the environment of using biofuels.

Excellent answers (5-6 marks)
Explain two advantages of using biofuels and state the major disadvantage regarding monocultures. Don't forget that both advantages and disadvantages need to be included in order to achieve full marks.

Biomass refers to organic matter (such as timber and crops) grown specifically to be burned in order to generate heat and power, or to be made into biofuels used in transport. For example, biogas is primarily made from methane, a gas from animal manure, while bioethanol and biodiesel are made from processing plant material from crops such as corn, sugar cane and rapeseed.

Biofuels for transportation

Fossil fuels have been used in transportation for many years, but they create pollution, emitting greenhouse gases that damage the atmosphere. They are also a finite resource, and the financial cost of diesel and petrol continues to rise.

The only realistic course of action for drivers and transport companies is to find less polluting and cheaper alternative fuels that address three key factors: good performance, reliability and widespread availability.

In principle, biofuels offer a way of reducing greenhouse gas emissions compared to conventional transport fuels. Currently, three types of commercially viable biofuels are available.

- **Biogas** is made from the fermentation of plant and animal matter. Waste materials that are used to make biogas include livestock manure, landfill site waste and sewerage sludge. Biogas production is a popular method of producing energy because it deals with the common problem of disposing of these waste materials.
- **Biodiesel** is made from monoculture crops (single crops grown on a large scale), including rapeseed, soybean and sunflowers. These crops are put through oilseed presses to extract the vegetable oil, which is used to make biodiesel.
- **Bioethanol** is made from natural sugar and starch from plants like sugar cane and sugar beet. Ethanol is an alcohol made through fermentation and distillation, a similar process to that used to produce wine and beer.

Biofuels are considered to be carbon neutral: they release the same amount of carbon when burned as the plants they are made from took in when they grew. This is because plants take in carbon dioxide in the process of photosynthesis so that they can grow.

From the environmental point of the view, the big issue with biofuels is biodiversity. With much of the Western world's farmland already given over to monocultures, the fear is that widespread use of biofuels will reduce habitat for animals and wild plants still further. If increased proportions of food crops such as corn or soy are used for fuel, this may push prices up, affecting food supplies for poorer countries.

Many people are putting their hopes into the second generation of biofuels, such as algae-based biofuel, which has the potential to make a significant contribution. The Carbon Trust believes that, by 2030, algae-based biofuel could replace over 70 billion litres of fossil fuels used in road and air transport.

Biofuel	Advantages	Disadvantages
biogas *Figure 6.8: The world's first biogas-powered train in Sweden*	• large amounts of waste biomass materials available from agricultural processes and landfill • diverts waste from landfill • burns more cleanly than fossil fuels • reduces the release of methane, a harmful greenhouse gas, into the atmosphere • by-products can be sold and used as compost and fertiliser to improve soil condition	• product value is low, making it an unattractive commercial product • low yield • process not very attractive economically compared to other biofuels on an industrial scale
biodiesel *Figure 6.9: Biodiesel pumps are already available across the country*	• can power regular diesel engines without need for engine modification • can be mixed with regular diesel to form biodiesel blends • could prolong the life of an engine as it lubricates and leaves fewer deposits • can be made at home using specialist equipment readily available • cleaner exhaust fumes – less carbon monoxide and soot and no sulphur dioxide like fossil fuels when burned	• ecological damage, including deforestation and intensive farming practices as a result of planting crops needed (monocultures) • monocultures reduce habitat variety • social and economic issues – crop prices increase, making them less affordable for developing countries • currently, expensive processing costs to convert biomass into fuels with low yield • incineration causes carbon dioxide pollution
bioethanol	• reduced harmful exhaust emissions – does not contribute to greenhouse effect • biodegradeable with no toxic effect on the environment • can be added to normal petrol to form ethanol-blended petrol • many engines can use ethanol without modifications • can be made at home relatively inexpensively, therefore reducing energy costs	
algae-based *Figure 6.10: Algae could be the future of biofuels*	most of the benefits above plus: • reduction of ecological damage • algae grows practically anywhere • crop does not affect freshwater sources or arable land • produces an oil which is relatively easy to convert to diesel and jet fuel • yield of vegetable oil from algae is comparatively much higher (about 30 times) than from land crops	• produces unstable biodiesel with many polyunsaturates • relatively new technology • performs poorly compared to biodiesel

Table 6.4: Advantages and disadvantages of biofuels for transportation

Climate change

Objectives

- **Understand** how climate change has been accelerated by increased greenhouse gases.
- **Understand** the responsibilities of 'developed' countries for reducing greenhouse gas emissions using the Kyoto Protocol.

Changes in climate have been happening naturally since the Earth was created 4.6 billion years ago. Natural changes in climate take place over long timescales. This slow rate of change has allowed species to adapt naturally to their changing surroundings. However, what we are witnessing now is an increase in this rate of change. A particular effect that has had a major impact on the Earth's surface temperature is the 'greenhouse effect'.

The greenhouse effect

Certain gases form a 'blanket' around the Earth and help insulate it from the escaping heat. The main gases that contribute to this 'greenhouse effect' are carbon dioxide, methane and nitrous oxide. This is a natural process, and without it the average temperature of the earth's surface would be about –18°C, instead of the more comfortable +15°C that it is today.

The natural greenhouse effect had maintained the Earth's temperature at around 15°C until about 200 years ago and the start of the Industrial Revolution. During this time, massive quantities of fossil fuels were burned, releasing large quantities of greenhouse gases into the atmosphere. More recently, the rate at which man-made greenhouse gases are being produced has accelerated dramatically, increasing the size of the 'blanket' around the Earth. More of the Sun's energy is now being trapped in the atmosphere, which in turn causes the Earth's temperature to increase more rapidly in a shorter period of time.

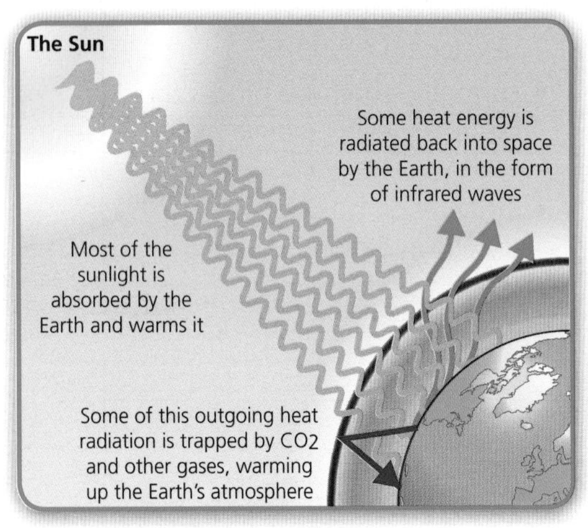

Fig 6.11: The 'greenhouse effect'

Carbon dioxide is the main man-made greenhouse gas. The energy we use in our homes produces 27 per cent of the carbon dioxide emissions in the UK, with another 25 per cent from domestic transport.

Some scientists estimate that a rise in global temperatures by 1.5 – 2°C could result in the polar ice caps melting, causing rising sea levels and flooding. Changes in weather patterns could produce extreme weather conditions, and expanded desert areas could displace whole populations.

The Kyoto Protocol

The Kyoto Protocol is an international agreement that sets targets for industrialised countries to cut their greenhouse gas emissions. It was adopted in 1997 in Kyoto, Japan, and came into force in 2005.

While184 countries have formally approved or ratified it, this important protocol acknowledges that developed countries are the ones mainly responsible for the current high levels of greenhouse gas emissions into the atmosphere. By signing up to the protocol, 37 industrialised countries and the European Community have committed to cut their combined emissions to five per cent below 1990 levels by 2008-2012; however each country has its own specific target, and more developed countries have higher targets to meet.

The world's most industrialised superpower, the United States has yet to ratify the Kyoto Protocol. Also, China and India, which are the world's largest polluters, were not included.

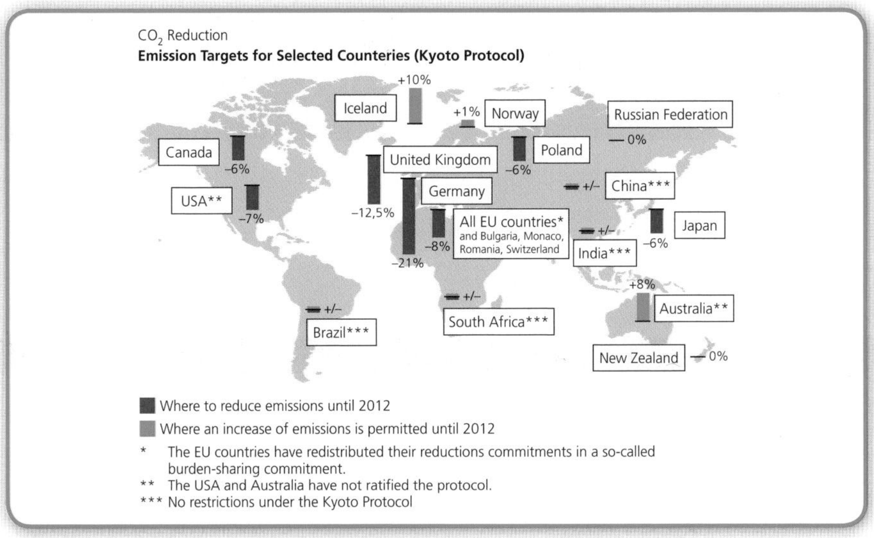

Figure 6.12: Emission targets for selected countries with the Kyoto Protocol

Problems with the Kyoto Protocol

The main problem with the Kyoto Protocol is trying to get every country to agree to cut their emissions in a relatively short timescale. The countries that showed early signs of meeting their goals, including Britain and Germany, had started working on energy-saving policy changes many years before the protocol became legally binding. There are other problems too: the USA feels it is unfair that China, its biggest trade competitor, can produce goods without any pollution restrictions, giving it the upper hand; and Canada, which does most of its trade with the USA, declared it would not be meeting its commitment.

The Kyoto Protocol is linked to the United Nations Framework Convention on Climate Change, but they are different: while the Convention *encouraged* industrialised countries to stabilise their emissions, the Protocol *commits* them to do so.

After Kyoto?

International post-Kyoto talks are now taking place to agree targets after 2012. The hope is that an early agreement will give enough time to prepare for the change, which will make success more likely the second time around. Further discussions will look at:

- **the creation of a global 'carbon market'** where companies can trade 'carbon credits'. A company that meets or exceeds its targets could sell carbon credits to another company. The idea is to bring emissions down on average by turning 'green' practices into moneymakers.
- **funding changes in developing countries**. For wealthy developed countries it is easier to be 'green', but poorer developing countries may not have the resources to change their production methods. A small percentage of the money spent by a developed country on a clean-energy project could be stored for developing countries to use to make changes.
- **the inclusion of rapidly industrialising countries**, such as China, which were not included in the protocol. A new agreement based on their polluting levels would cancel some other countries' targets.

Support Activity

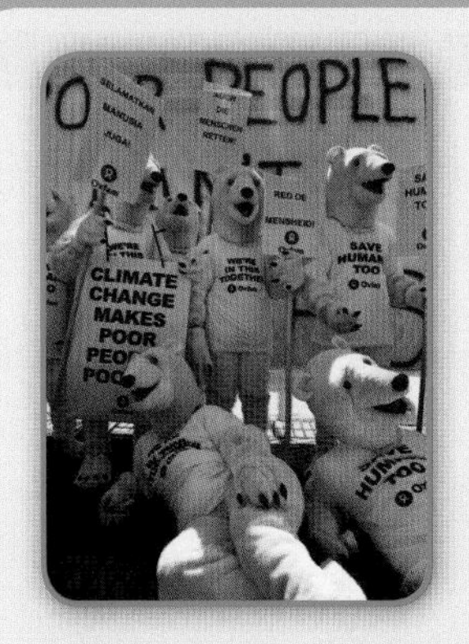

Here are some protesters at international climate change talks. Why are they dressed as polar bears? Explain your answer using issues relating to climate change and the Kyoto Protocol.

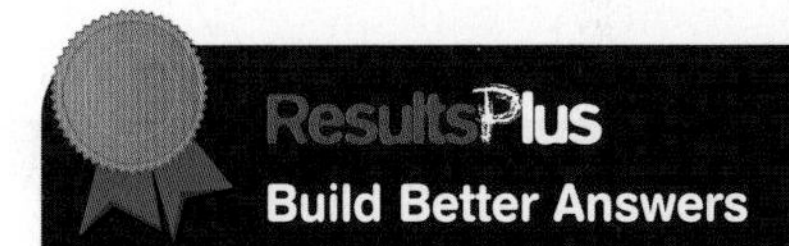

Discuss the global environmental issues that require the Kyoto Protocol. (6 marks)

Basic answers (1-2 marks)
Identify one issue, with little understanding of its environmental impact.

Good answers (3-4 marks)
Identify several issues, with some understanding of their environmental impact.

Excellent answers (5-6 marks)
Identify a range of relevant issues, with a detailed understanding of their environmental impact.

Know Zone
Chapter 6 Sustainability

Sustainability means safeguarding the world for ourselves and for future generations. Sustainable product design involves our need to design and manufacture products using energy and materials in a way that minimises the depletion of finite resources, waste production and pollution. These are not simply issues for governments and large companies – how can you contribute towards a more sustainable future?

You should know...

the following about sustainability:

Sustainability	Principals	Characteristics	Uses	Advantages and disadvantages	Responsibilities
Reducing	✓	×	✓	✓	×
Reuse	✓	×	✓	✓	×
Recovering	✓	×	✓	✓	×
Recycling	✓	×	✓	✓	×
Wind energy	×	✓	✓	✓	×
Solar energy	×	✓	✓	✓	×
Biomass/biofuels	×	✓	✓	✓	×
Global warming	✓	×	×	×	✓
Kyoto Protocol	✓	×	×	×	✓

Key terms

Questions relating to sustainability will require you to know the principals of minimising waste production, using renewable sources of energy and how governments have a responsibility to address concerns about global warming.

Property	Meaning
Principles	The reasons for minimising waste production using the 4 Rs and reducing CO_2 emissions using the Kyoto Protocol.
Characteristics	Recognisable features that help to identify or differentiate one process from another. For example, why are renewable sources better for the environment than non-renewable sources when generating energy?
Advantages and disadvantages	Qualities and features favourable to success or failure. For example, what are the advantages of using laser cutting rather than traditional cutting processes?
Responsibilities	The duty and obligations of developed countries to prevent/slow down the rate of global warming.

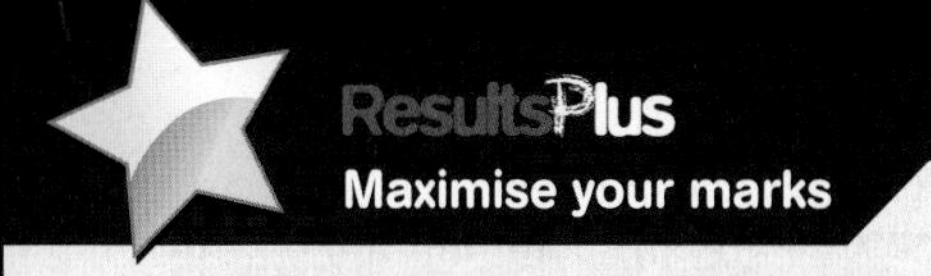

Grade E-C range question: Explain **one** benefit to the environment of using traditional glass milk bottles from a 'door-to-door' delivery service. (2 marks)

Student answer	Examiner comments	Build a better answer
Glass milk bottles can be washed out and used again (1 mark) *which makes the milk cheaper.* (0 mark)	'Re-using' glass milk bottles is an appropriate benefit to the environment but the justification is neither correct nor a benefit to the environment.	Glass milk bottles can be washed out and reused many times (1 mark). Saving the raw materials and energy required to produce a new one. (1 mark)

Overall comment: Be careful to make sure that your justification for an 'explain' question clearly relates to the point you have made. You may not get the second mark if the examiner feels it is totally unrelated.

Grade C-A range question: Justify the use of wind as a source of energy for generating electricity. (4 marks)

Student answer	Examiner comments	Build a better answer
Wind turbines are a clean way of producing electricity for the National Grid (1 mark) *at a relatively low cost* (1 mark). *They can be used on a large scale in wind farms especially off the UK coast harnessing off-shore winds.* (1 mark)	A good response that clearly demonstrates an understanding of wind energy's potential in the UK. However, they have missed the opportunity to state the obvious that wind energy is non-polluting and sustainable compared to fossil fuels.	Using wind turbines is non-polluting as it does not require burning fossil fuels (1 mark) and sustainable as wind is a renewable resource (1 mark). In large-scale wind farms, in mountainous and off-shore regions with strong winds, (1 mark) it can produce low-cost power. (1 mark)

Overall comment: It is a real shame that this student did not go on to make one final point and achieve full marks. Re-read your responses and pick out the key points yourself if needed, adding an extra point.

'Stretch and Challenge' A/ A* question: Discuss the problems associated with recycling schemes as organised by local councils. (6 marks)

Student answer	Examiner comments	Build a better answer
Local councils sometimes have problems with people not sorting their rubbish properly (1 mark) *which means it's difficult to collect the right type of materials* (1 mark). *Some people don't take the environment too seriously and don't recycle at all* (1 mark) *and the council can't do much about it.* (1 mark)	This student's response focuses upon people not being receptive to local council recycling schemes which is perfectly fine. However, the issues go beyond just people's attitudes and these need to be addressed in order to achieve full marks.	It is expensive for councils to operate local recycling schemes (1 mark) as collection days, vehicles and man-power are needed (1 mark). Some houses do not sort all their waste by polymers, glass, metals, paper and organic waste (1 mark) so some collections are hard to recycle (1 mark). Some households are reluctant to recycle (1 mark) and the council has few powers to force them (1 mark). Many modern products are not easily recycled (1 mark) as they are made from different materials e.g. a PET bottle with a HDPE cap. (1 mark)

Overall comment: Consider these extended-writing questions carefully before you put pen to paper. You could even plan your response on a scrap piece of paper. Do not include your planning on the actual exam paper.

Chapter 7 Ethical design and manufacture
Moral, social and cultural issues

Objectives

- **Understand** the concept of a 'throwaway' culture.
- **Understand** the concept of built-in obsolescence.
- **Describe** the forms of built-in obsolescence in new products.

Figure 7.1: In fast-food restaurants, consumers expect their food to be served on demand and in convenient, disposable packaging

Support Activity

1. Make a list of products that you are familiar with (and perhaps have bought yourself) under the four headings of built-in obsolescence.
2. How could the physical obsolescence of ink cartridges for printers be solved? You may want to refer to the section on minimising waste production.

Our 'throwaway' culture

The Industrial Revolution of the eighteenth century fundamentally changed life in many countries. Large-scale mechanisation led to changes in production, workforce and transportation. Population explosions occurred in towns and cities where production was now centred, creating a new, urban way of life.

More people needed more products, and mass production responded to this need. Expensive and time-consuming handcrafted work could now be replaced by machine work. Products once made exclusively for the rich could now be made in large numbers, and at a price that ordinary working people could afford.

Our modern consumer society is a feature of the rich, developed world, where people's wants are satisfied by a continual supply of new and relatively inexpensive products. This human society, which is strongly influenced by mass-consumerism, is often referred to as a 'throwaway' culture. It features the over-consumption and excessive production of short-lived or disposable products.

Built-in obsolescence

Built-in or planned obsolescence is a method of creating consumer demand by designing products that wear out or become outmoded after limited use.

In the 1930s, an enterprising engineer working for General Electric proposed increasing sales of torch bulbs by increasing their efficiency and shortening their lifespan. Instead of lasting through three batteries, he suggested that each bulb last only as long as one battery. By the 1950s, built-in obsolescence had been widely adopted by a range of industries, most notably in the American motor and domestic appliance sectors.

Modern companies still use built-in obsolescence because it means that consumers have to keep buying a product, either because it wears out or because it becomes unfashionable or out-of-date. This benefits the company because it keeps sales up and maintains profits.

Modern companies currently use four forms of built-in obsolescence: technological, postponed, physical and style (see Table 7.1).

Form of obsolescence	Description	Examples
technological	companies are forced to introduce new products with increased technological features as rapidly as possible to stay ahead of the competition occurs mainly in the computer and electronics industries	sales of multimedia mobile phones such as the iPhone or Blackberry have increased in comparison with normal mobile phones because of their many functions, including social networking
postponed	companies launch a new product even though they have the technology to realise a better product at the time	products such as laptop computers are obsolete almost as soon as they are purchased; a newer version with a better specification will soon be introduced
physical	when the very design of a product determines its lifespan	disposable or consumable items such as light bulbs, one-trip disposable packaging, ink cartridges for printers
style	changes in fashion mean that products seem out of date, and customers want to replace them with current, 'trendy' goods	high-street fashion collections; premiership club kit which is updated every season, so fans need to constantly purchase new replica football kits

Table 7.1: Forms of built-in obsolescence

Sustainable product design

Sustainable product design may be the solution to built-in obsolescence and the 'throwaway' culture. It aims to produce products that have a low impact on climate change and the depletion of the world's resources.

Today, when protecting the environment is such a priority, the question of product life and durability is a concern. Consumers are applying more and more pressure on companies to address these issues through more ethical design. Designers must now consider how products can be upgraded, reused, recycled or maintained for extended product life, rather than simply being thrown away.

Stretch Activity

Discuss the advantages and disadvantages of being a consumer in a 'throwaway' culture. Here are some things to think about.

- What would you miss if they were no longer readily available?
- What are the environmental issues relating to over-consumption?

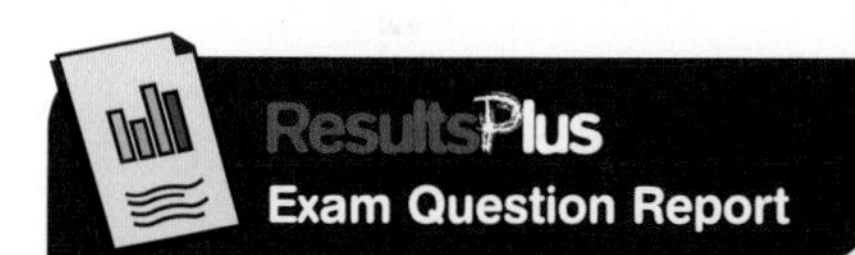

Explain two environmental disadvantages of built-in obsolescence. (4 marks)

How students answered

Most students only stated one environmental disadvantage such as 'more products thrown away', with no real justification.

71% 0-1 marks

Many students were able to state two environmental disadvantages, and some successfully justified one of their responses.

24% 2-3 marks

Some students could fully justify two disadvantages of built-in obsolescence.

5% 4 marks

Objectives

- **Understand** the concept of offshore manufacturing.
- **Explain** the advantages and disadvantages of offshore manufacturing in developing countries.

Support Activity

Make a list of products that are manufactured abroad by British companies. Here's one for starters: Dyson vacuum cleaners – but where are they based now?

Multinationals

The need to be competitive means that many companies sell their products all over the world. A multinational is a company that does business in more than one country. Multinationals are usually based in developed countries, but have offices, branches or manufacturing plants in other countries. Such companies can be so powerful that their budgets exceed the gross domestic product (GDP) of many developing countries.

Offshore manufacturing

Offshore manufacturing means manufacturing products in a different country, usually for export back to the original country to be sold. For example, a UK clothing company might run a factory in Asia, and then sell the products back in the UK.

Offshore manufacture is a driving force in today's global marketplace, with multinationals increasingly using it as a strategic tool. Many multinationals draw on the expertise of other countries to develop new products, especially in the field of technology. Others are relocating to less developed countries such as India, China or former Soviet nations and outsourcing their work. Modern corporate headquarters and industrial estates are sprouting up in these countries to supply the new demand for offshore manufacturing. Initially jobs in developing countries were created through the manufacture of shoes, cheap electronics and toys; more recently, many simple customer service operations such as call centres have also moved to these countries.

Offshore manufacture works by using the internet and high-speed data networks that cover the entire globe. Design data can simply be sent to another country for manufacture: for example, a book can be designed in the UK but sent via digital networks to China for printing.

Why do multinationals manufacture offshore?

The reason multinational manufacture offshore is quite simple: it costs them less. It is now possible to achieve the same quality work at a fraction of what it would cost if Western companies manufactured in their own countries. For example, mould-making for injection moulding is generally much more affordable in China than in the West (about 50 per cent lower in China, or 30 per cent lower in Taiwan). By having bases in developing countries, multinationals can also get better access to expanding overseas markets.

Figure 7.2: Some multinationals have been accused of operating 'sweatshops' with low wages and poor working conditions for their employees

Advantages	Disadvantages
• economic regeneration of local areas through increased employment • improved living standards through potential for career development and multi-skilling of local workforce • physical regeneration of local area through development of road system, transportation and local amenities • widening of the country's economic base and enabling of foreign currency to be brought into the country • enables the transfer of new technology that would be impossible without the backing of multinationals	environmental issues: • increased pollution and waste production as a result of large-scale manufacturing activities • destruction of local environment to build factories, processing plants, roads, etc. employment issues: • lower wages than those of workers in developed countries, where a minimum wage operates • promotion restrictions as managerial roles are often occupied by workers from the multinational's home country • no unions to protest against equal rights issues, such as unfair dismissals • lower health and safety standards when using sweat shops • devaluing of traditional craft skills and replacement by repetitive tasks • local communities can become dependent on multinational leaving them devastated if the multinational leaves the area

Table 7.2: Advantages and disadvantages of offshore manufacturing for developing countries

Is it ethical?

Offshore manufacture raises a number of ethical issues, including its impact on unemployment in developed countries, and the exploitation of labour in developing countries. Why would a British-based multinational company continue to pay the minimum wage to its UK employees when they could employ Indian or Chinese labour for 50-60 per cent less? But is it right that workers in developing countries may not be given the opportunities for promotion, pay rises, company benefits, union membership and working conditions that their Western colleagues demand as basic human rights? As multinationals build centres of operation in new areas, the local workforce may be displaced from their traditional trades, becoming dependent on the largely unskilled labour that many industrial processes require.

Stretch Activity

Discuss the advantages and disadvantages of offshore manufacturing for:

- multinational companies
- workers in developing countries
- workers in developed countries.

ResultsPlus
Build Better Answers

Evaluate the use of offshore manufacturing by multinational companies. (6 marks)

Basic answers (0-2 marks)
Give up to two advantages to multinationals of using offshore manufacturing.

Good answers (3-4 marks)
Explain two advantages of using offshore manufacturing to multinationals and the local economy.

Excellent answers (5-6 marks)
Explain two advantages for multinational companies but also explain one disadvantage regarding employment issues. Don't forget that both advantages and disadvantages need to be included in order to achieve full marks.

Objectives

- **Understand** that some brand names may cause offence to different cultures.
- **Understand** that designers and multinationals must use local design and local experts in a global market.

Cross-cultural design

Culture is the behaviour of human beings acting as groups and as individuals in a community. Culture plays an important role in the design of products, and cross-cultural design is especially important in the global market. Products designed for one culture may not fit another, so there cannot be a one-size-fits-all approach. Designers need a better understanding and tolerance of the diversity of different cultures.

Global marketing

When marketing globally, it is important to consider what different cultures will think of a brand. The wrong translation of phrases or the wrong perception of a logo or brand identity can damage potential growth in a new country and can often offend customers.

- 'Barf' is an Iranian laundry detergent. 'Barf' means 'snow' in Farsi
- 'Fart' bars are candy from Eastern Europe
- 'Pee Cola' from Ghana

Multinationals operating in a global market are increasingly using local designers and experts who understand local cultures. 'Localisation' is the process of developing a product to operate successfully in a new culture. Local experts can test and evaluate the product to see how it will best fit into the target market, and can provide feedback that is invaluable in developing the design further.

The importance of studying different cultures is perhaps best shown when multinationals get it wrong. Here are two examples of how companies failed to understand the target market when launching their products in India.

Soggy cornflakes

'Kellogg's set up a branch in India and started producing cornflakes... What they didn't realise was that Indians, rather like the Chinese, think that to start the day with something cold, like cold milk, on your cereal is a shock to the system,' observed Indian cultural critic Homi Bhabha. 'And if you pour warm milk on Kellogg's Corn Flakes, they instantly turn into wet paper.'

Kellogg's made the mistake of assuming that people in India started their day in the same way as people in Britain or America. Products created for the American or European market are not necessarily relevant to a user in India or Africa where the conditions, customers and thinking are different. Basic product concepts may need to be completely redesigned. In this case, Kellogg's ended up pulling their cornflakes from shelves and re-developing them so that they could stand up to warm milk.

The only sure way for companies to avoid mistakes is to research the local market thoroughly. Design teams need to look for cultural differences, both obvious and more subtle ones. They may need to look into a country's history, religious beliefs, climate, geography, languages, ideas of beauty and, sometimes, popular culture. If companies get it wrong, there can be multi-million-dollar consequences.

The washing machine ate my sari

As part of an aggressive global marketing strategy, Whirlpool designed a stripped-down washing machine for developing countries. Named the 'World Washer', it was launched in Brazil, Mexico, China and India, with slight design and styling modifications for each market to reflect local tastes. For example, the 'delicates' program was re-labelled 'sari cycle' on the Indian model.

The washing machine ended up doing very well for the company everywhere but in India. With tens of millions of dollars at risk, Whirlpool sent a team to India to find out what had gone wrong. They finally realised that the garments people were washing were little more than sheets of fine cotton or silk, six to nine metres long, such as saris. These garments were getting tangled up and shredded in a millimetre-wide gap between the machine's moving parts. That single millimetre forced Whirlpool to completely restructure their marketing strategy for India, as well as to design a new washing machine. It took the company years to recoup their losses and regain a significant market share in the country.

Because Whirlpool designers did not fully understand specific target markets, the World Washer failed to live up to its name. The basic mistake Whirlpool made was to assume that needs are the same across all markets. It had a limited understanding and little direct experience of the customs and styles of dress in India.

Cautionary tales like those of Kellogg's and Whirlpool have prompted an increasing number of multinationals to use local expertise to develop appropriate products. For example, Unilever has established a network of more than 68 'innovation centres' in 20 countries worldwide.

Figure 7.3: Understanding local culture is extremely important if designers want to develop successful products

ResultsPlus
Build Better Answers

Explain two reasons why a company needs to consider cross-cultural design when marketing a new product worldwide. (4 marks)

Basic answers (0-1 marks)
State one generic reason for considering cross-cultural design issues such as 'so you don't offend anyone', with no justification.

Good answers (2-3 marks)
State up to two good reasons, but only fully justify one of them.

Excellent answers (4 marks)
State and fully justify two important reasons why companies consider cross-cultural issues when marketing their products in other countries.

Support Activity

1 Why would you probably not be able to easily use a product such as a computer in China or Saudi Arabia?

2 Can you think of any products or brands that you use that would have to be re-developed for an international market?

Stretch Activity

Discuss why a multinational company might use local expertise to develop products in countries worldwide.

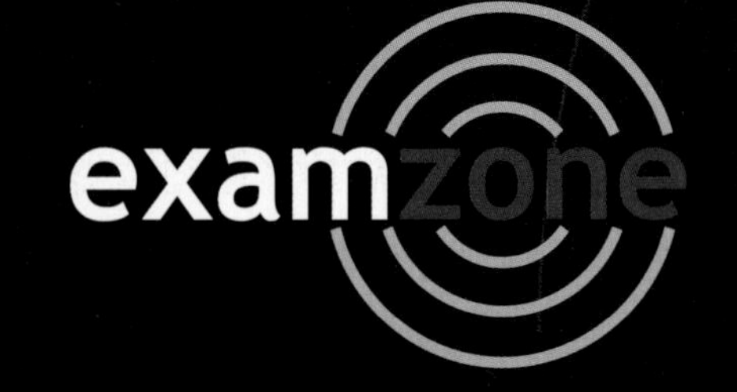

Know Zone
Chapter 7 Ethical design and manufacture

As a designer, you need to know about the properties of a wide range of materials and components so that you can make informed choices about their use in certain products.

You should know...

about the following about ethnical design and manufacture:

Technology	Strategy	Characteristics	Advantages and disadvantages	Uses
Built-in obsolescence	✓	✓	✓	✓
Offshore manufacturing	✓	✓	✓	✓
Cross-cultural design	✓	✓	✓	✓

Key terms

Questions relating to ethical design and manufacture will require you to know the methods (both good and bad) that companies use to market their products world-wide.

Property	Meaning
Strategy	A plan or method for achieving a specific goal or result. For example, a marketing strategy for selling products in international markets.
Characteristics	Recognisable features that help to identify or differentiate one process from another. For example, what are the different types of built-in obsolescence?
Advantages and disadvantages	Qualities and features favourable to success or failure. For example, why is offshore manufacturing good for multinationals but has its problems for the local workforce?

ResultsPlus
Maximise your marks

Grade E-C range question: Explain **one** advantage, to a multinational company, of using offshore manufacturing. (2 marks)

Student answer	Examiner comments	Build a better answer
The company will make products in another country because they are cheaper to make. (1 mark)	A good attempt but the student does not fully justify why manufacturing in another country is cheaper.	It is cheaper to manufacture a product in a developing country (1mark) because the average wage for workers is much lower than in developed countries. (1 mark)

Overall comment: A question like this requires you to fully understand the topic and use specialist terminology in your response i.e. developing and developed countries.

Grade C-A range question: Outline the reasons why cross-cultural design is important when marketing a new product world-wide. (4marks)

Student answer 	Examiner comments	Build a better answer 
If you don't design a product with other cultures in mind it may fail in one country (1 mark). *This is because it may be offensive to some people* (1 mark). *Some electrical products won't work in other countries because they use different power supplies and sockets.* (1 mark)	This response starts off well with the student making a relevant point and justifying it. However, the next point about 'different power supplies' is going off the point a bit but is awarded a 'benefit of the doubt' mark as it is a suitable example.	Cross-cultural design is important when launching a product world-wide because a company needs to know local markets (1 mark) by using expert knowledge in that country who know the culture better than anyone (1 mark). If companies don't thoroughly research local needs and wants then the product could fail (1 mark) which could cost them millions of dollars in redevelopment costs. (1 mark)

Overall comment: It is always worth having an 'educated guess' if you can't think of anything more to write. However, that doesn't mean you should 'waffle' on with no clear direction as this is just a waste of your precious time in the exam.

'Stretch and Challenge' A/ A* question: Evaluate the use of 'built-in obsolescence' as a marketing strategy for a new product. (6 marks)

Student answer  (achieving 4 mark)	Examiner comments	Build a better answer  (achieving 6 mark)
Built-in obsolescence is a cynical marketing ploy by companies to make sure you keep buying their products as they wear out quickly (1 mark). *Products could be built to last but that would mean you wouldn't need a new one and this loses companies money* (1 mark). *For example, if you buy a new games console there will already be another one ready to go which has a better technology* (1 mark). *In fashion you have to keep up with the latest trends in order to fit in even though your clothes are still wearable.* (1 mark)	This student has made some relevant points but they are all negative aspects of built-in obsolescence. Advantages of built-in obsolescence to companies also need to be addressed. The use of specific examples is fine because it is clear to see that the student is referring to 'technological' and 'style' forms of obsolescence.	Built-in obsolescence is used in products because it means that you have to keep buying a product because it wears out (1 mark). This is good for a company because it keeps sales up and therefore profits. (1 mark) However, there are many negative aspects for the consumer which include: • Electronic products are upgraded too often so one becomes obsolete too quickly (1 mark); • Fashion moves on each season and perfectly good clothes are not worn because they are considered unfashionable (1 mark); • Consumables such as ink cartridges are sealed and simply thrown away when they could be re-filled (1 mark); • Some products are not built to last on purpose so that they have to be replaced (1 mark).

Overall comment: Your response to an 'evaluate' question has to address both advantages and disadvantages in order to achieve full marks. For example, you could outline five disadvantages and just one advantage. The use of a bullet pointed list is particularly useful in this instance and clear for the examiner to read.

Your controlled assessment

In Unit 1 Creative Design and Make Activities, you will complete a design activity and a make activity based on one of a range of tasks provided by the exam board, Edexcel. The design and make activities can be linked (combined design and make) or separate (design one product, and make another).

How much is it worth?

Unit 1 Creative Design and Make Activities is worth 60 per cent of your overall GCSE Graphic Products course. This is a major unit, so you should always try to produce your best work for all stages of the creative process.

What is a graphic product?

Edexcel Graphic Products has two clearly defined pathways: either conceptual design or the built environment.

1. Conceptual design incorporates a wide range of 3D products with associated graphics, such as packaging design, product/industrial design, point-of-sale (POS) display and vehicle design.
2. The built environment focuses on the man-made surroundings that provide the setting for human activity, such as architecture, interior design, exhibition design, theatre sets and garden design.

What is controlled assessment?

Controlled assessment means that your work will be carried out under 'controls' you present to ensure that all of the work is your own. Details of the controls for the design activity are on page 111, and for the make activity are on page 139. Task setting is one control.

What is task setting?

Edexcel will provide five 'tasks' or broad themes for you to choose from. The examples given are just suggestions: You can choose your own projects providing they fit under one of the broad themes. Following a separate design and make route gives you greater flexibility. For example:

- your design activity could use the 'packaging' task and you could design the blister pack for a new toy
- your make activity could use the 'interior and architectural design' task and you could make a scale model of your bedroom.

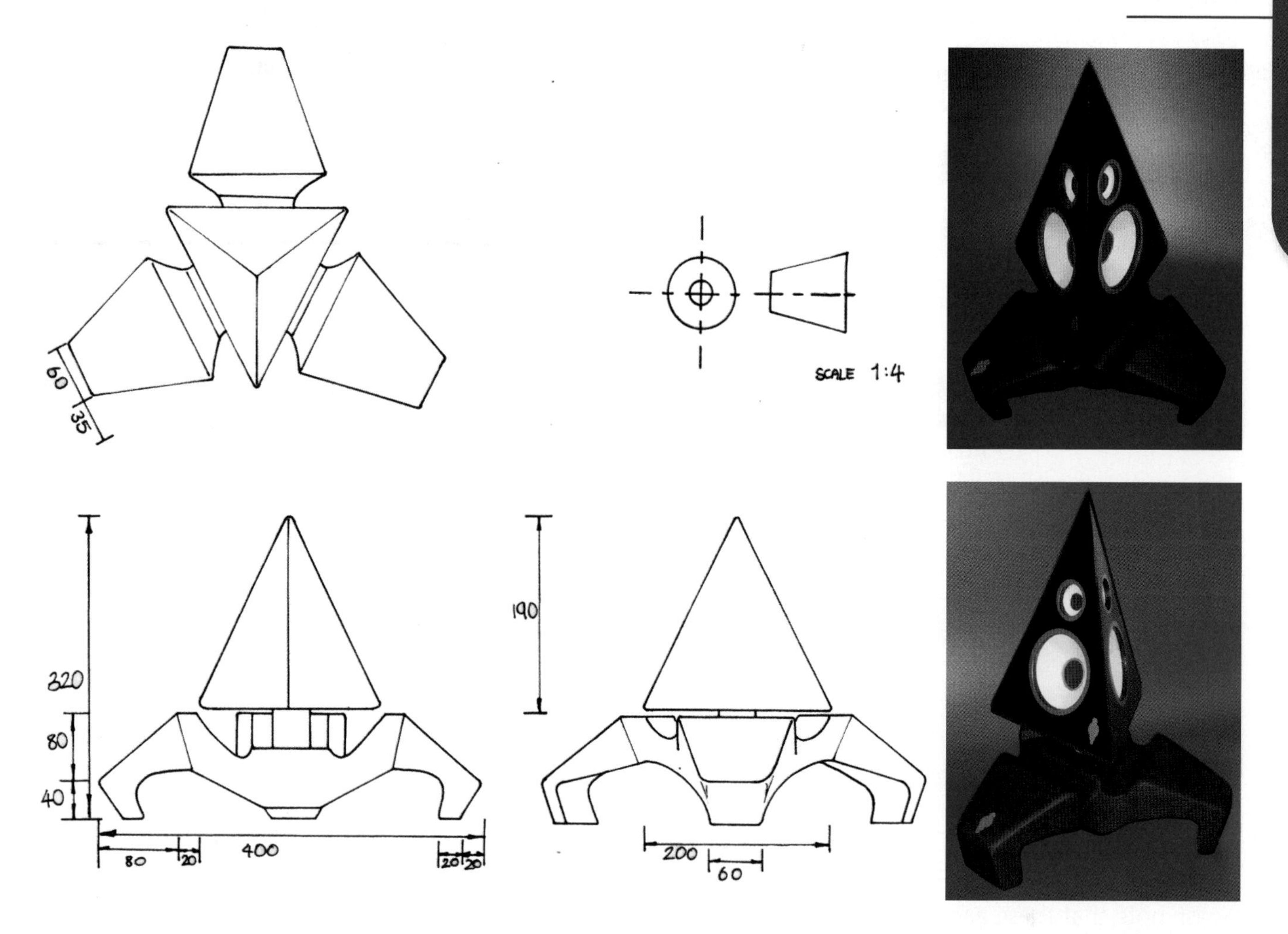

How will the work be assessed?

Your design and make activities will be assessed by your teacher according to the criteria in Table 8.1, and you will be awarded a mark out of 100.

Table 8.1: Stages and criteria for assessment of design and make activities

Design activity (50 marks) You will undertake a design activity covering the following three stages and eight assessment criteria:		**Make activity (50 marks)** You will undertake a make activity covering the following three stages and five assessment criteria:	
Stage 1: Investigate (15 marks)		**Stage 4: Plan (6 marks)**	
1.1	Analysing the brief (3 marks)	**4.1**	Production plan (6 marks)
1.2	Research (6 marks)		
1.3	Specification (6 marks)		
Stage 2: Design (20 marks)		**Stage 5: Make (38 marks)**	
2.1	Initial ideas (12 marks)	**5.1**	Quality of manufacture (24 marks)
2.2	Review (4 marks)	**5.2**	Quality of outcome (12 marks)
2.3	Communication (4 marks)	**5.3**	Health and safety (2 marks)
Stage 3: Develop (15 marks)		**Stage 6: Test and evaluate (6 marks)**	
3.1	Development (9 marks)	**6.1**	Testing and evaluation (6 marks)
3.2	Final design (6 marks)		

A sample from your group will be sent to Edexcel to check that your teacher's marks are accurate.

Chapter 8 Design activity

Introduction

You will undertake a design activity covering the three stages and eight assessment criteria listed in Table 8.2. Table 8.2 gives you an idea of the time you should allow for your work to meet each criterion, and of the number of pages you might produce for each.

Table 8.2: Outline of the design activity

Stage	Assessment criterion	Marks	Suggested times	Suggested pages
1. Investigate	1.1 Analysing the brief	3	1 hour	1
	1.2 Research	6	3 hours	2–3
	1.3 Specification	6	1 hour	1
2. Design	2.1 Initial ideas	12	5–6 hours	2–3
	2.2 Review	4	1 hour	1
	2.3 Communication	4	Evidenced throughout design and development stages	N/A
3. Develop	3.1 Development	9	5–6 hours	2–3
	3.2 Final design	6	1–2 hours	1–2
	TOTAL	50	17–20 hours	10–14

Please note that these are only suggested times and numbers of pages for each assessment criterion – they are not compulsory. However, we strongly recommend that you keep to the deadline of 20 hours for the whole of this design activity. You should be able to achieve high marks for each assessment criterion within the suggested number of pages. The exam board is looking for 'quality rather than quantity', so one page packed with relevant information or detailed designs is better than several pages with irrelevant or inadequate information or lots of blank space.

Starting points

There are several starting points to this activity. For example, you can devise your own design brief from a topic that you are particularly interested in, or your teacher can provide one. The important thing to remember is that your design brief must fit under one of the five broad themes stated in the Edexcel set tasks for that particular year.

You will also need to consider whether you are going to undertake:

- combined design and make activities (where you will make what you design)
- separate design and make activities (where you design one product and make an entirely different one).

A separate design activity may give you the opportunity to undertake a 'blue sky' project, where you can let your imagination run and demonstrate your designing skills, creativity and innovation by developing a product for the future.

Examples: Future communications

Here are two examples that would fit well under the Edexcel set task heading 'Concept design'.

Wireless food scanners to change the way we shop

Bob has a nut allergy and has to be careful what he eats. In the past, this made food shopping time-consuming, as he had to examine each food packet carefully to check that the product was free from any nut traces.

Bob now has a portable nutritional content scanner. This allows him to scan food wrappers, which have been fitted with Radio Frequency Identification (RFID) technology, to identify quickly and easily whether it contains nuts. If Bob is unsure whether a product is suitable, he can contact his doctor, via his home hub, to ask for an expert opinion.

Smart drug dispensing and on-body monitors

Parul is mildly asthmatic and always carries a smart inhaler with her. She gets automatic reminders on her home hub and hand-held communicator about factors such as atmospheric conditions that may make her asthma worse, as well as advice on how to cope.

The number of times she uses the inhaler is automatically monitored and she is contacted if her usage is above or below the norm. Parul has also signed up for an anywhere/anytime service that allows her to meet other local people with asthma and provides general advice on managing her condition. She has access to a named care adviser and coordinator, and could attend a clinic, either real or virtual, if she wanted to.

Class Activity

Read the two starting points on future communications. In small groups, discuss and answer these three questions.

1 What are the key words/issues that need to be explored in greater depth? Brainstorm your ideas using a spider diagram.

2 What would these products look like? Would they be designed to be familiar, like other hand-held devices? Sketch some initial ideas.

3 How would these products feel? How could they be designed to be user-friendly? Sketch some technical details.

Now answer the following question on your own.

4 Design a wireless food scanner or a smart drug dispenser as a mini project.

Controls

Design activity

Task setting

- Edexcel will provide five tasks for you to choose from.

Preparation

- You can undertake research and preparatory work outside the classroom without supervision.

Write-up

- You must complete the write-up of your portfolio under informal classroom supervision.
- You may use ICT to word-process some of your write-up, provided that this work has been drafted in the classroom under direct supervision and that the final version matches the original draft.

Stage 1 Investigate (15 marks)

Objectives

- **Analyse** your design brief in enough detail to be able to clarify design needs.
- **Analyse** the key words and phrases that help you understand the issues related to your design brief.

Controls

Investigate

- Your design brief must be based on a task set by the exam board, Edexcel.
- You can research your design brief at home or in the library and bring your notes into the class.
- The write-up of your analysis must be completed in a classroom.
- Your teacher can provide you with feedback to make sure you have analysed your design brief sufficiently.

ResultsPlus
Watch out!

If you use concise and succinct analysis, you should be able to fit this work onto a single A3 sheet. Avoid unnecessary information and 'padding'.

1.1 Analysing the brief

To get top marks you will:
write a precise statement of what you are going to design (and make), in the form of a design brief
analyse key words relating to your design brief
provide sufficient information to enable you to research your project in greater depth.

Design brief

Your design activity should start with a design brief that fits under one of the Edexcel set task headings. The design brief should be a precise statement of what it is you are going to design, and perhaps make. Here is an example.

Separate design and make activities

Task: Point-of-sale display

Design brief: To design a point-of-sale display for a new music magazine aimed at teenagers. The point-of-sale display should include a space for the magazine cover to be changed each month when a new issue comes out. I am also going to design the front cover of the first issue of the magazine.

Combined design and make activities

Task: Interior design and architecture

Design brief: To design and make a scale model of a new social area at The Ravensbourne School for Year 11 learners to relax, study and eat in.

You do not have to use a real client for your project but, if you have the opportunity to, they could provide invaluable feedback at several stages of the design process. You must, however, identify an appropriate user group/target market group (TMG).

Analysis

Here you need to find a route through your project by identifying the main issues that need to be addressed and focusing on what exactly needs to be done. There are many ways of analysing your design brief including brainstorming, mood boards (collages of images and key statements) and attribute analysis. However, it is extremely important that you use any or all of these to focus on a specific design problem, rather than in a general way.

Brainstorming

You can do this alone or with a small group. Write the design brief in the middle of a large piece of paper, then note important things to consider before you start designing. These subsections can be broken down further to help you think about the problem in more detail.

Mood boards

You must identify a user group or target market for your product. This will help you to be selective with your research and will aid your designing. You can create a mood board relating to your target market, to set the mood for the project. You do not need a separate page to do this.

Attribute analysis

Attribute analysis is useful for clarifying key features of a design brief. Start by drawing up a table with relevant headings: for example, possible materials and possible size. Then list all the possibilities in each column, and highlight the ones that apply to your project. This will help you to form statements to help your design.

Example analysis work

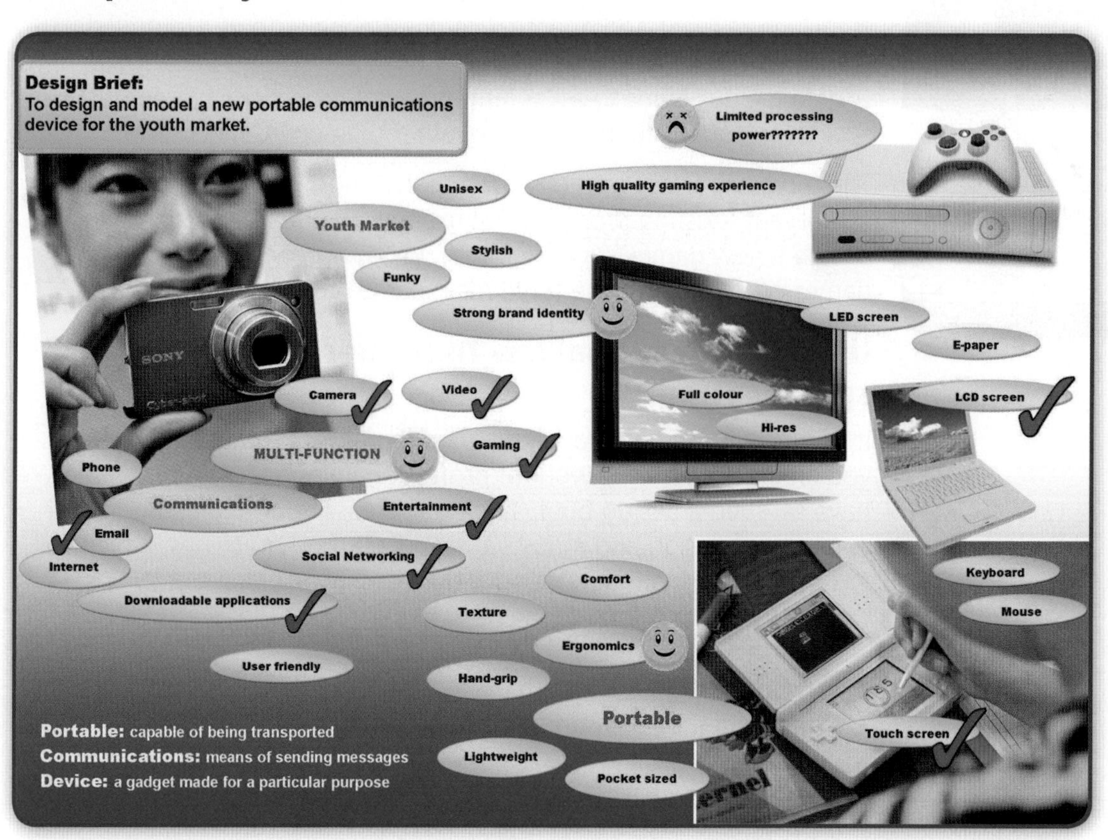

Moderator's comments

This learner has presented their analysis clearly and concisely. The design brief is clearly stated on the page and several methods have been used to analyse key issues including, brainstorming and mood boards. This page sets the scene well for the project.

Objectives

- **Present** selective and focused research that addresses the needs you identified when analysing your brief.
- **Analyse** relevant, existing products in terms of their performance, materials components, processes, quality and sustainability issues.
- **Apply** the findings from your product analysis and research to inform the writing of your own specification criteria.

- The main focus of your research must be product analysis, not mood boards or contrived questionnaires.

1.2 Research

To get top marks you will:
use selective and focused research
analyse at least one existing and relevant product in detail including its performance, materials, components, processes, quality and sustainability issues
determine sustainability issues using life cycle assessment (LCA)
determine key factors from your product analysis that you will have to include in the specification for your own product.

Selective and focused research

It is really important that your research is selective and avoids repetition and unnecessary padding. Here are some ways to avoid this.

- When gathering information from the internet or a book, select the relevant information only and represent the key points on your page. Avoid simply accumulating pages of unnecessary information.
- Avoid using mood boards here as they will not gain you any marks. You can a mood board as part of your target market group analysis (see pages 112–113), provided it is fully annotated.
- Include relevant and quantitative questionnaires only. Contrived questionnaires are a waste of controlled assessment time. If you have genuine questions to ask a target market group, use closed questions with categories to choose from rather than open-ended ones.

The most important part of your research must come from the analysis of relevant, existing products. For example, if a learner is going to design a new games controller, they will obviously analyse existing games controllers for the three main games consoles.

Using product analysis to influence your specification

It is important that you analyse existing graphic products to give you an insight into the work of professional designers and how they have satisfied a design brief. This will increase your understanding of commercial design and industrial manufacturing processes, which will in turn inform and influence your own design and make activities.

Here are some important questions to ask about a product:

- What are the technical considerations that affect its performance?
- What materials and components are used and why?
- What manufacturing processes are used and why?
- What issues of quality are involved in its design and manufacture?
- What are the sustainability issues throughout its life cycle?

You can try to find the answers to these questions by using several methods of analysis including:

- disassembly of a product(s), and
- comparison of similar products using common criteria.

Example research work: disassembly

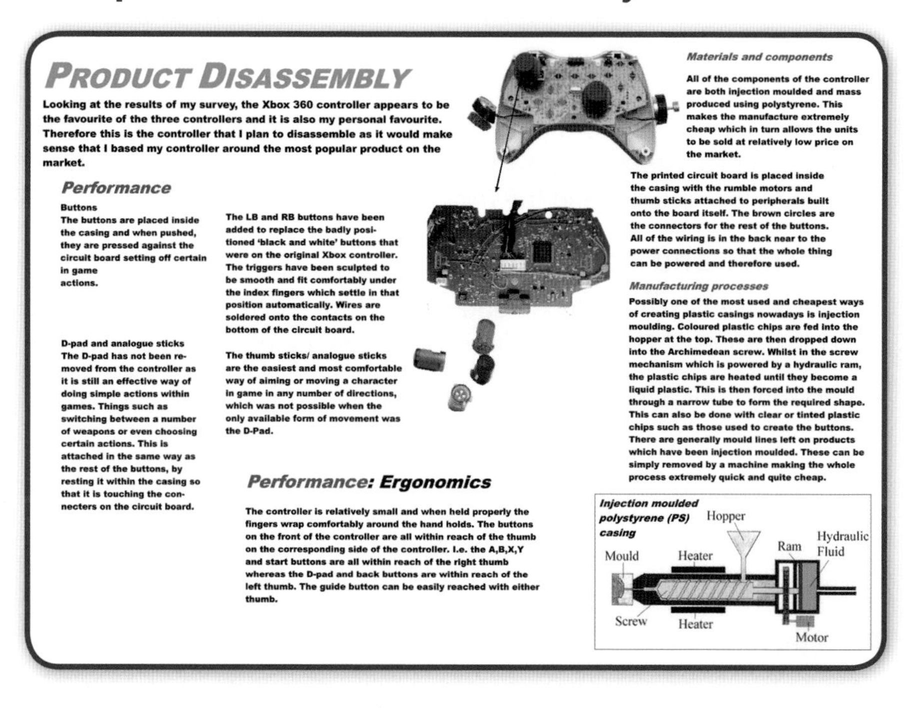

Moderator's comments

Disasassembly of an existing product is extremely useful in determining your specification. This learner has taken apart a games controller in order to show the materials, components and processes used to make it. The learner can test the ergonomic factors and can test other performance requirements through game-play. This approach enables the learner to better understand what needs to be considered when writing their specification and designing their own games controller.

Example research work: comparison

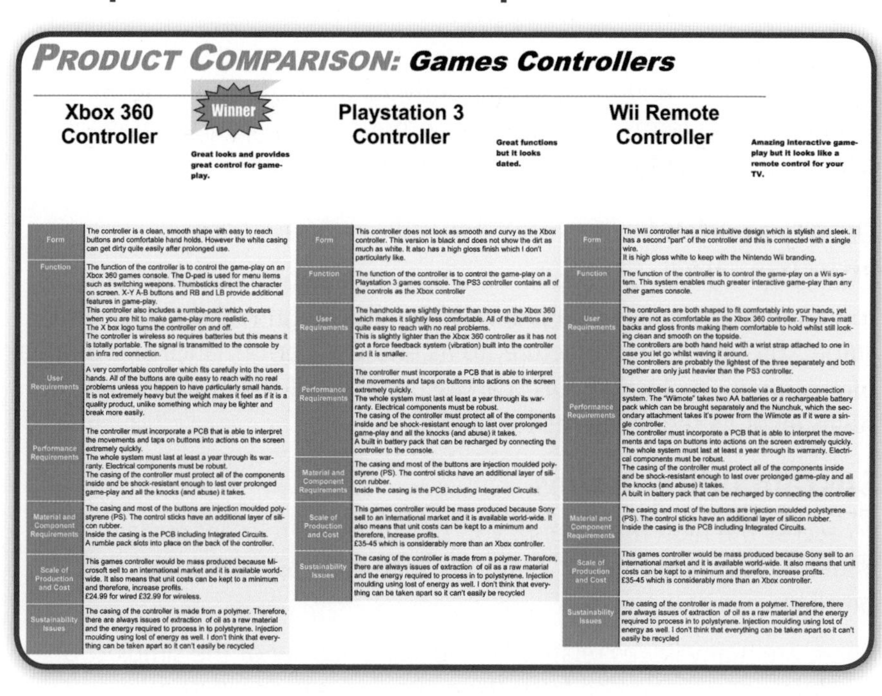

Moderator's comments

This learner has used the basic product specification from Unit 2 as common criteria. The controllers differ in user and performance requirements. This is a great way of determining which product works best and should be used as a piece of 'good practice' when designing your own product.

Product analysis

Performance

The performance of a product is determined by the choice of materials, components and processes used to make specific parts of it. For example, performance factors for opening a drinks can could include the size of the ring pull and the thickness of the aluminium.

Materials and components

The correct choice of materials and components is crucial if a product is to perform well. Designers will consider a range of factors including:

- limitations – if a material is not 'fit for purpose' it will fail, such as folding box board used for packaging heavy products.
- wear and deterioration – some materials will naturally wear and deteriorate such as newpapers but if you need a product to last a long time, you must use materials that will resist wear and not deteriorate.

Processes

Modern production processes have greatly increased the design values of many mass-produced products. One good example is the process of injection moulding polymers, which has influenced the design of the casings of electrical goods and packaging (see page 45).

Quality

When designing products, quality refers to the product's ability to satisfy a need and, more importantly, its fitness for purpose. In manufacturing terms, quality means producing a product that is the best it can be, fully functional and free from defects.

Quality assurance

Quality assurance (QA) systems are put in place by a manufacturer to monitor the quality of a product from its design and development stage, through its manufacture, to its end-use performance and degree of customer satisfaction. QA is an assurance that the end product fulfils all of its requirements for quality.

Sustainability issues

To access the high mark band, you must feature relevant sustainability issues in your research: for example, a life cycle assessment of the product.

There are many issues relating to the sustainability of the simplest of products, as you can see from the example on the next page.

Example product analysis work: performance

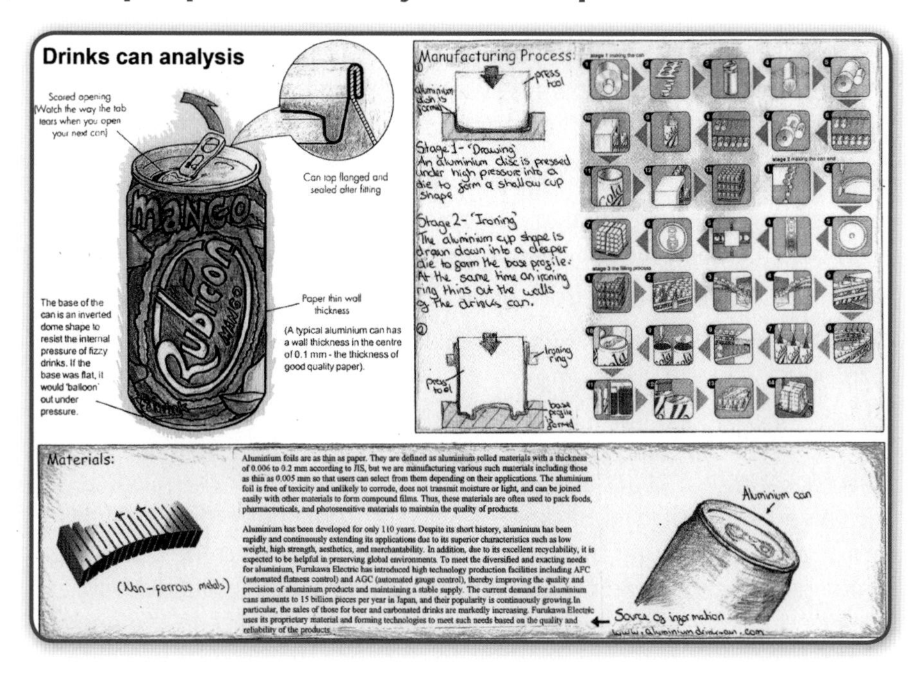

Moderator's comments

Here, a learner has analysed a drinks can by focusing on the materials, components and processes used to manufacture the mass-produced product. Performance issues are labelled on the drinks can itself.

Example product analysis work: sustainability

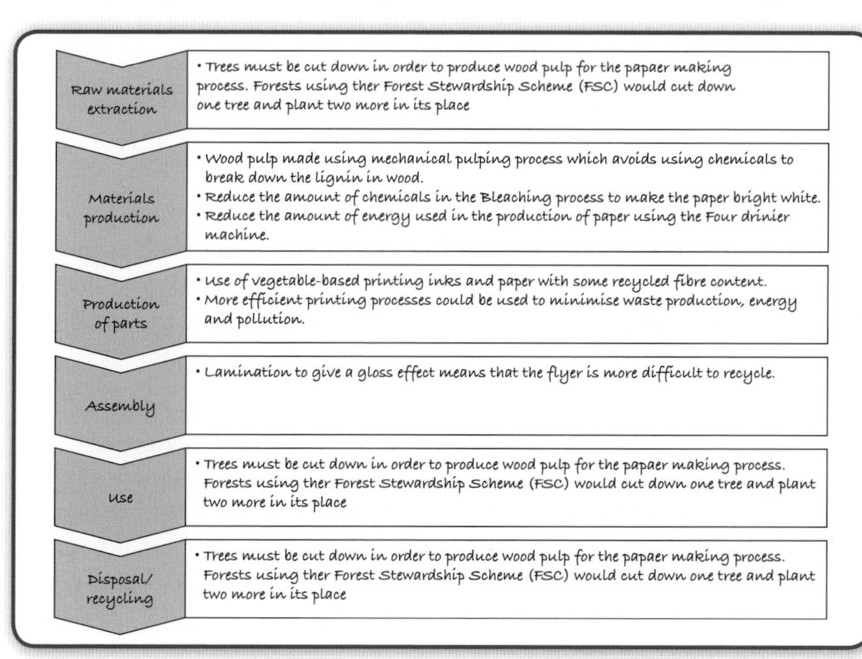

Raw materials extraction	• Trees must be cut down in order to produce wood pulp for the papaer making process. Forests using ther Forest Stewardship Scheme (FSC) would cut down one tree and plant two more in its place
Materials production	• Wood pulp made using mechanical pulping process which avoids using chemicals to break down the lignin in wood. • Reduce the amount of chemicals in the Bleaching process to make the paper bright white. • Reduce the amount of energy used in the production of paper using the Four drinier machine.
Production of parts	• Use of vegetable-based printing inks and paper with some recycled fibre content. • More efficient printing processes could be used to minimise waste production, energy and pollution.
Assembly	• Lamination to give a gloss effect means that the flyer is more difficult to recycle.
Use	• Trees must be cut down in order to produce wood pulp for the papaer making process. Forests using ther Forest Stewardship Scheme (FSC) would cut down one tree and plant two more in its place
Disposal/ recycling	• Trees must be cut down in order to produce wood pulp for the papaer making process. Forests using ther Forest Stewardship Scheme (FSC) would cut down one tree and plant two more in its place

Moderator's comments

This learner has used a life cycle assessment well to determine sustainability issues when analysing a club flyer.

Objectives

- **Understand** the four main functions of packaging.
- **Understand** the reasons for choosing appropriate materials for different uses.

Analysing packaging

The role of packaging

The role of packaging is extremely important to consider when analysing any product. Packaging has four main functions:

- to contain the product inside and dispense it when necessary
- to advertise the product contained inside using brand identity and inform the customer of the contents (e.g. food labelling)
- to protect the product from breakages and spillages
- to preserve the contents to prevent them from spoiling (e.g. food products).

A container

The package should act as a carrier and a dispenser for the product contained within it. The package has to transport the product safely from the manufacturer, first to the retailer and finally to the consumer. The type of package and the materials used in its construction will directly influence the effectiveness of this function. Once in the hands of the consumer, the container has to dispense the product in a safe and convenient way. This can be achieved by a variety of methods including lids, twist caps, aerosols or tear strips.

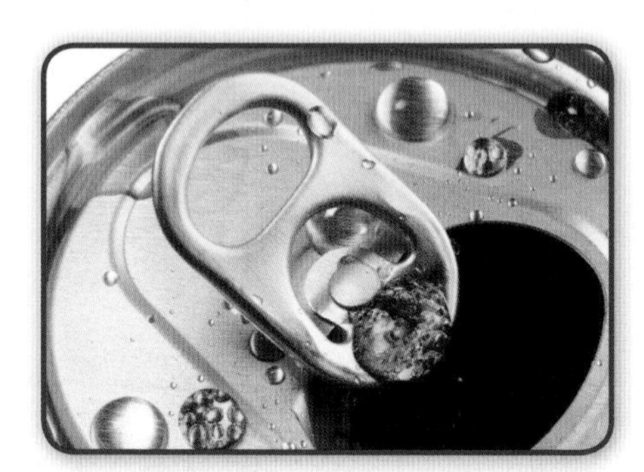

Figure 8.1: Aluminium cans are often used to contain a fizzy drink, with a ring-pull to open

What your analysis could include

- How does the package contain the product? For example, what type of package is it (e.g. blister package)? What materials have been used (e.g. packaging laminate)?
- How does the package dispense the product? For example, what type of closure does it use (e.g. lid, cap)?
- Does the package contain the product effectively? For example, are there any problems with opening and closing the package?

An advertiser

The package should both describe the product and identify it by means of brand identity. The consumer should know what it is that he or she is buying in terms of its contents, its weight and how the product should be used, including any special instructions.

Figure 8.2: Soap powder boxes are designed to 'shout' from the supermarket shelves

Manufacturers will use a brand identity to sell the product. Most products will use printed graphics to make the consumer aware of its presence on the shelf. Decades of advertising and brand identity persuade the consumer to buy the product because the brands have become familiar and trusted. Faced with a wide choice of similar products, the well-known brand name has consistent quality and the consumer can rely on it. Here, the package can make the choice of product for the shopper.

What your analysis could include

- Does the package contain all the necessary labelling and easily understood instructions for use (e.g. food labelling regulations)?
- How does the package differentiate itself from its competitors (e.g. strong colour scheme and brand identity)?

A protector

The transportation of a product without breakage or spillage is of prime importance, and it is the role of packaging to ensure this. The choice of material for a particular product has serious consequences for the effectiveness of the package. For example, glass bottles are an excellent package for fizzy drinks but they are easily broken if accidentally dropped. Some packages, such as biscuits containers, have internal packaging to protect their brittle contents. The amount of packaging material used can cause environmental problems with disposal after use.

What your analysis could include

- How do the materials used in the package offer protection to the product? For example, how does a corrugated board box with expanded polystyrene inserts protect the contents?
- Could alternative materials be used? For example, could PET be used instead of glass?
- Could less packaging material have been used? For example, does the product have to be individually wrapped?

A preserver

The package must be able to preserve the product for the necessary amount of time to prevent spoilage. Some products have greater shelf lives than others: for example, fresh vegetables may be packaged in a tray covered with plastic film as they have a shelf life of a couple of weeks, while tinned vegetables can keep for months or years due to the cooking and canning process.

What your analysis could include

- How does the package preserve the product? For example, which processes and materials are used (e.g. blow-moulded PET bottle to preserve the 'fizz' in a soft drink)?

Figure 8.4: Preserving foods with different shelf lives

Support Activity

1 Analyse a product that you are familiar with using the four main functions of packaging as criteria.

2 Discuss examples of packaging that do not meet these four main functions. What are the major problems in their use and how could these be overcome?

Stretch Activity

Use the internet to research Easter Egg packaging. Use the '4 Rs' from Chapter 2 to analyse their packaging and suggest ways of minimising waste production.

Figure 8.3: Easter eggs can be very attractive but over-packaged

Objectives

- **Understand** what makes an effective brand and logo.
- **Understand** the importance of brand identity in creating the 'right image'.

Analysing brand identity

Modern life is full of branded goods, from the cornflakes we eat to the trainers we wear. Our choice of brands says a lot about us as individuals. All graphic products have a brand identity and it is important that we analyse this in order for us to design our own.

Logos, symbols and trademarks

The term 'branding' comes from the practice of marking property such as cattle with a hot iron to prove ownership. Today, the most recognisable feature of brand is a distinctive name, logo, symbol or trademark.

Logo	Description	Example
Iconic/symbolic	These use simple icons and symbols to represent a particular company or product. They create an image that is instantly recognisable and memorable without the need for text.	
Logotype/wordmark	These present a company or brand name in a uniquely styled typeface. Different typefaces give different impressions to the intended audience. Images can also be integrated into a logotype to create great visual impact.	DELL
Combination marks	These are graphics with both text and symbols/icons. The visual impact of the symbol or icon is complemented by simple text, which identifies the company or product still further.	Domino's Pizza®

Table 8.3: Types of logo

Typefaces

The choice of typeface is important in giving the right image for the product. On a computer, typefaces are referred to as fonts. There are four main categories: serif, sans serif, script and decorative.

Category	Examples	Characteristics	Image
Serif (typeface has 'tails' at end of letters)	Times New Roman, Courier, Century, Garamond	Easy to read and pleasing on the eye	Traditional
Sans serif (typeface has no 'tails')	Arial, Calibri, Tahoma, **Impact**	Strong, bold and clear	Modern
Script	Comic Sans, *Brush Script*, *Forte*, *Mistral*	Looks personal (handwritten) but can be difficult to read	Historical/ personal
Decorative	Jokerman, Chiller, **STENCIL**, Ravie	Attracts attention but some can be difficult to read	Modern (graffiti style) or historical (Old English) depending on style

Table 8.4. The four main categories of typeface

Colour

Logos should usually use a limited range of colours. The most effective logos use a maximum of three colours, but the choice of just which colours is important for the brand.

BRITISH AIRWAYS

Figure 8.5 :The British Airways logo uses the red, white and blue colours of the British national flag

In focus: Nike

The best symbols are simple and distinctive and manage to say something about the nature and quality of the products or services. One of the most famous sports brands is Nike. The company takes its name from the winged Goddess of Victory, and the distinctive Swoosh logo is a simplified version of one of her wings. Nike uses several variations of its logo but the Swoosh is now so recognisable that the company does not even have to include the word Nike anymore!

What your analysis could include

- What type of logo is it?
- What typeface does it use?
- How many colours does it use, and which ones?
- What image does the logo create?

Creating the right image

The concept of branded goods is central to our society. Many people use brands as a means of identification with others, as a signal to others of status or personal values. Advertising encourages us to buy branded goods to buy into a particular lifestyle. For example, a person driving a sports car with a BMW badge on the bonnet will automatically be branded successful and wealthy.

Analysing brand image

For any brand to be successful it must portray the right image. When analysing the image of any product ask yourself:

- who is it aimed at (the target market)?
- what does it say about the people who buy it (their lifestyle)?

The Nike drawstring bag

The Nike drawstring bag is a common sight. This simple nylon bag with a drawstring closure is used to carry sports kit, books and other items, and is aimed primarily at the youth market. The Swoosh logo sends the message that the wearer is aware of fashion and quality. Other bags carry other logos, but do not send out the same message.

Support Activity

Picture yourself walking into a sports shop. Make a list of the first three brands you look out for. Now explain why you have chosen these three brands, in this order.

Stretch Activity

Discuss whether it is right that people often want the more expensive branded product rather than a less expensive product that does a similar job.

Objectives

- **Understand** the stages in disassembling a graphic product.
- **Understand** the materials and components, and the manufacturing and printing processes, involved in the production of a graphic product.

Figure 8.7: Frijj chocolate milkshake

Product disassembly

You can get a clearer understanding of how a product is manufactured and put together by taking apart an existing product and examining its various parts or components. This will help you to work out:

- the function of its components
- the materials used
- the manufacturing processes used
- the printing processes used.

Worked example: Milkshake bottle

There are many flavoured milkshakes available on the market. We are going to look at a bottle of Frijj chocolate milkshake. First, we will take the bottle apart in order to determine its component parts.

Stage 1: Working out the function of component parts

Here we can clearly see that the milkshake bottle is made up of four parts: a cap, a seal, a plastic bottle and a plastic label. Now we must work out what each part does.

- The bottle cap provides a means of opening the bottle in order to dispense (pour) the drink.
- The seal provides a means of ensuring that the drink is kept fresh and provides security so that the contents cannot be tampered with.
- The bottle is a container for the drink.
- The label has the brand identity printed on it for easy product recognition and also carries all of the relevant legal labelling.

Stage 2: Working out the materials used

Each component part has been made of a specific material that will ensure it performs as intended. We know that the cap, bottle and label are made of a polymer, but which polymer exactly? By looking for the coding symbol on each part, we can check which type of polymer has been used and determine the properties and reasons for use.

- The cap is made from low density polyethylene (LDPE 4) because it is tough and hard-wearing: the cap includes a screw thread which will be opened and closed several times during its use. This polymer will not flavour the liquid inside. Note: LDPE is not covered on this course so a little bit of further research was required.
- The seal is part of the cap and so is also made from LDPE. This polymer is very flexible which means that the seal can be torn to open.

- The bottle is made from polyethylene terephthalate (PET 1) because it is very tough, does not flavour the liquid inside and provides an excellent barrier to protect the milkshake from contamination.

The polymer film label has no visible code printed on it, so we must wait until we determine its method of manufacture and then we might have some further clues as to what it is made of.

Stage 3: Working out the manufacturing processes used

By applying our knowledge and understanding of thermoforming processes, we can work out the method of manufacture for each component part.

- The bottle itself is hollow, so we know it has been blow-moulded. It has a faint line running down each side and across the bottom. From this we can work out that the bottle has been formed by a split mould (two halves).
- The bottle cap and seal are not hollow but have been formed into a dish shape. We can therefore work out that the cap has been injection moulded.
- The polymer film label fits tightly around the bottle, which is a characteristic of shrink wrapping. We would have to further research shrink wrapping to fully understand the process and the types of polymers used.

Results of further research

Shrink wrapping using PVC sleeves offers all-round graphic coverage. The bottle is labelled when full. A PVC sheet is rolled and heat-sealed to form a tube. The PVC tube is placed over the bottle and an infrared hot air system applies heat (pre-heated to 58°C). The polyvinyl chloride (PVC) sleeve shrinks at between 60°C, 90°C, moulding itself around the bottle. The bottle is rotated during the process to ensure an even shrinkage all around.

Stage 4: Working out the printing processes used

The PVC shrink sleeve has been printed on to provide the brand identity and advertise the product.

By applying our knowledge and understanding of commercial printing processes, we can work out that the PVC shrink sleeve was printed using either the gravure or the flexographic process. Flexography is currently the most widespread of the two printing processes, so it is more likely that this process was used.

Apply it!

You can analyse any graphic product by applying your knowledge and understanding from Unit 2, undertaking further research and a little bit of educated guesswork.

Support Activity

1. Why was a shrink-wrapped label used for this product? Explain your answer.
2. How would the design of the bottle have to be changed so that a flat label could be used instead of a shrink-wrapped label? Explain in detail.

Stretch Activity

Following the stages for disassembling a product carefully, disassemble another product of your choice.

Objectives

- **Produce** realistic, technical, measurable specification points that address some issues of sustainability for your own product.
- **Justify** your specification points using findings from your research.

ResultsPlus
Watch out!

Each specification point needs to be a full justification, and not simply a statement. This is similar to the way in which an 'explain' type question in your Unit 2 exam needs a statement and then a justification to get full marks.

1.3 Specification

To get top marks you will:
write specification points under the seven headings outlined below
include issues of sustainability relating to your product in your specification
justify each specification point
apply findings from your research to the specification criteria.

The specification is one of the most important documents in your project as it will determine the criteria for your finished design. It will form the basis of all your evaluative comments during the design and development stage, and of the testing and evaluation of the finished product at the end of the project.

The design specification is a list of bullet points that are more specific than the design brief. It must cover all of the important aspects relevant to your design brief.

It would be beneficial to use the headings from Chapter 4 Designing products in Unit 2 of the specification.

- **Form** – how should your product be shaped/styled?
- **Function** – what is the purpose of your product?
- **User requirements** – what qualities would make your product attractive to potential users?
- **Performance requirements** – what are the technical considerations that you must achieve within your product?
- **Material and component requirements** – how should materials and components perform within your product?
- **Scale of production and cost** – how will your design allow for scale of production and what are the considerations in determining its cost?
- **Sustainability** – how will your design allow for environmental considerations?

Don't forget to apply the findings of your research when writing your specification. For example, when disassembling a fizzy drink bottle, you should have identified the most suitable materials and processes (for example, blow-moulded PET for the bottle). You should now use these findings in the specification for your new fizzy drink bottle.

Example specification work

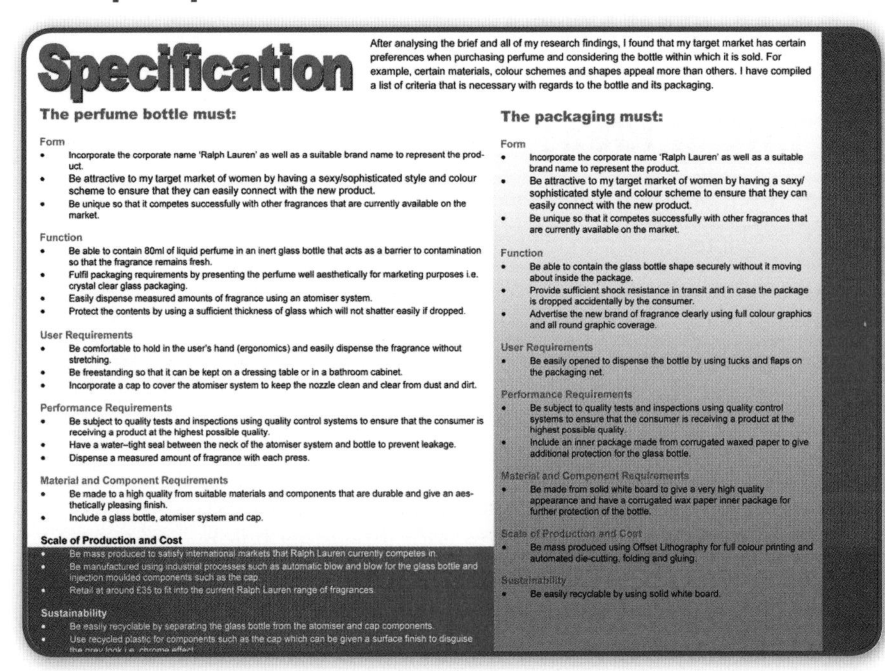

Specification

After analysing the brief and all of my research findings, I found that my target market has certain preferences when purchasing perfume and considering the bottle within which it is sold. For example, certain materials, colour schemes and shapes appeal more than others. I have compiled a list of criteria that is necessary with regards to the bottle and its packaging.

The perfume bottle must:

Form
- Incorporate the corporate name 'Ralph Lauren' as well as a suitable brand name to represent the product.
- Be attractive to my target market of women by having a sexy/sophisticated style and colour scheme to ensure that they can easily connect with the new product.
- Be unique so that it competes successfully with other fragrances that are currently available on the market.

Function
- Be able to contain 80ml of liquid perfume in an inert glass bottle that acts as a barrier to contamination so that the fragrance remains fresh.
- Fulfil packaging requirements by presenting the perfume well aesthetically for marketing purposes i.e. crystal clear glass packaging.
- Easily dispense measured amounts of fragrance using an atomiser system.
- Protect the contents by using a sufficient thickness of glass which will not shatter easily if dropped.

User Requirements
- Be comfortable to hold in the user's hand (ergonomics) and easily dispense the fragrance without stretching.
- Be freestanding so that it can be kept on a dressing table or in a bathroom cabinet.
- Incorporate a cap to cover the atomiser system to keep the nozzle clean and clear from dust and dirt.

Performance Requirements
- Be subject to quality tests and inspections using quality control systems to ensure that the consumer is receiving a product at the highest possible quality.
- Have a water–tight seal between the neck of the atomiser system and bottle to prevent leakage.
- Dispense a measured amount of fragrance with each press.

Material and Component Requirements
- Be made to a high quality from suitable materials and components that are durable and give an aesthetically pleasing finish.
- Include a glass bottle, atomiser system and cap.

Scale of Production and Cost
- Be mass produced to satisfy international markets that Ralph Lauren currently competes in.
- Be manufactured using industrial processes such as automatic blow and blow for the glass bottle and injection moulded components such as the cap.
- Retail at around £35 to fit into the current Ralph Lauren range of fragrances.

Sustainability
- Be easily recyclable by separating the glass bottle from the atomiser and cap components.
- Use recycled plastic for components such as the cap which can be given a surface finish to disguise

The packaging must:

Form
- Incorporate the corporate name 'Ralph Lauren' as well as a suitable brand name to represent the product.
- Be attractive to my target market of women by having a sexy/sophisticated style and colour scheme to ensure that they can easily connect with the new product.
- Be unique so that it competes successfully with other fragrances that are currently available on the market.

Function
- Be able to contain the glass bottle shape securely without it moving about inside the package.
- Provide sufficient shock resistance in transit and in case the package is dropped accidentally by the consumer.
- Advertise the new brand of fragrance clearly using full colour graphics and all round graphic coverage.

User Requirements
- Be easily opened to dispense the bottle by using tucks and flaps on the packaging net.

Performance Requirements
- Be subject to quality tests and inspections using quality control systems to ensure that the consumer is receiving a product at the highest possible quality.
- Include an inner package made from corrugated waxed paper to give additional protection for the glass bottle.

Material and Component Requirements
- Be made from solid white board to give a very high quality appearance and have a corrugated wax paper inner package for further protection of the bottle.

Scale of Production and Cost
- Be mass produced using Offset Lithography for full colour printing and automated die-cutting, folding and gluing.

Sustainability
- Be easily recyclable by using solid white board.

Moderator's comments

This learner has decided to design both the perfume bottle and its packaging. This combined approach works for products like this but, if you were undertaking an interior design or architectural project, there would be no need to add more outcomes, as they are big enough projects as they are.

Realistic, technical and measurable

Here are some examples of realistic, technical and measurable points.

Realistic

- Be comfortable to hold in the user's hand (ergonomics) and easily dispense the fragrance without stretching.

This specification point is realistic because the design and development using 3D modelling will be able to determine an ergonomically sound bottle shape that can be easily used. It can be physically tested by the user group in order to gain valuable feedback.

Technical

- Be able to contain 80 ml of liquid perfume in an inert glass bottle that acts as a barrier to contamination so that the fragrance remains fresh.

This specification point is technical because it outlines a specific quantity of fragrance that the bottle must hold. It then goes on to specify glass as the most appropriate material for containing the fragrance as a result of the learner's research.

Measurable

- Be attractive to my target market of women by having a sexy/sophisticated style and colour scheme to ensure that they can easily connect with the new product.

This specification point is measurable because the learner can quite easily ask the target market group if they like the style of the new product.

Stage 2 Design (20 marks)

Objectives

- **Present** alternative initial design ideas that are realistic, workable and detailed.
- **Demonstrate** your understanding of materials, processes and techniques applicable to your initial design ideas.
- **Apply** your research findings to your initial design ideas.
- **Address** specification points through your initial design ideas.

Controls

Design

- You can produce rough sketches of your ideas outside the classroom and bring them into the class.
- Your initial ideas must be copied up into best under supervision in a classroom.
- Your teacher can provide you with feedback to make sure that you have produced enough design ideas and that they are sufficiently annotated.

ResultsPlus
Watch out!

You don't have to design both a 2D and a 3D graphic product; you can concentrate on one or the other. However, some projects suit a combined approach, such as a perfume bottle with its packaging and brand identity.

2.1 Initial ideas

To get top marks you will:
sketch 3–5 different design ideas for your product(s)
indicate all the relevant materials, processes and techniques that would be required to make the product commercially
apply your research findings to your designs or include additional research where appropriate
annotate each design with reference to relevant specification criteria.

Generating initial ideas is an important point in the design process. This is where you can demonstrate your individual flair by creating alternative ideas that fulfil your specification criteria. Your ideas should focus on key specification points, and your annotation should explain how these are met, as well as why the materials and processes would be appropriate.

Use graphical techniques that you have developed over your Graphic Products course and which you are comfortable with. The most important thing is to communicate your ideas as best as you can. Try to incorporate different materials and processes into your design ideas to show an understanding of more than one material and process.

Presenting your ideas

There are two main approaches, both of which are acceptable.

Graphic thinking

Some people find it easy to translate what they are thinking directly into rough sketches. Design sheets look busy, with lots of overlapping sketches. You can highlight the most important designs using subtle shading or by fully rendering them. The other sketches will still be there, but will stay more in the background.

More formal drawings

Other people like to practise their ideas in sketchbooks, then communicate them as more polished drawings. These sheets look more formal with the majority of designs in full colour.

Whichever technique you use, make sure that all your drawings are fully annotated to show your knowledge and understanding of relevant materials and processes. Additional drawings showing parts of the design in more detail are extremely useful. Most importantly of all, make sure that your design ideas relate directly back to your specification points.

Include new 'snippets' of information when the exploration of a design idea requires the learner to look at specific issues in greater detail.

Example ideas work: graphic thinking

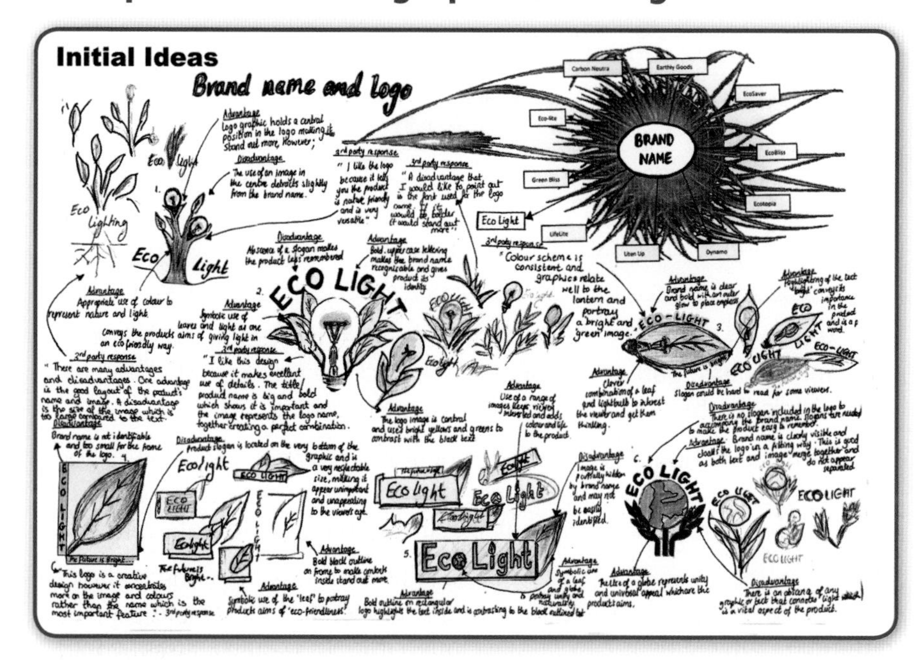

Moderator's comments

Good design sheets are 'busy' and include 'graphic thinking', featuring fully annotated sketches that communicate different design intentions in detail. The exam board does not really expect the use of CAD to feature strongly at this stage; what they are looking for is the exploration of ideas, and not their refinement.

ResultsPlus
Watch out!

Due to the time constraints imposed by controlled assessment, it would be a good idea for you to cut back on excessive decoration of your design sheets. It is the design ideas themselves that are the star of the show, not over-embellished borders and backgrounds.

Example ideas work: more formal drawing

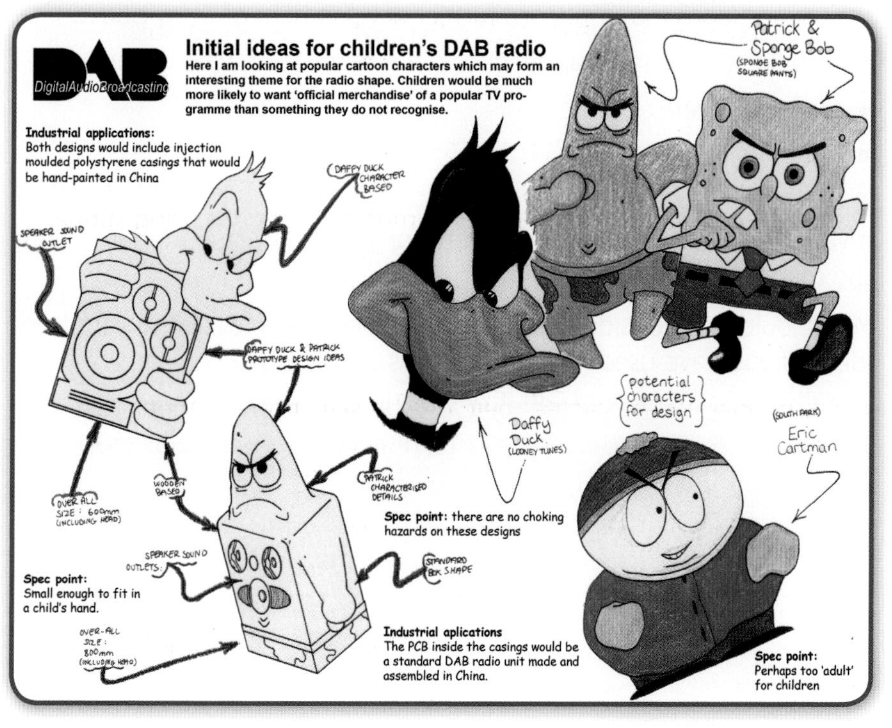

Moderator's comments

This learner has used a more formal method of communicating their initial design ideas. Research into children's themes is applied to the initial ideas extremely well. Annotation is sufficient to communicate information about which materials and processes would be used for the mass-produced product, and relates back to several specification points.

Objectives

- **Present** objective evaluative comments against your original specification criteria.
- **Evaluate** your initial design ideas using user-group feedback and issues of sustainability.

2.2 Review

To get top marks you will:
test at least two of your best ideas against relevant points in your original specification criteria
evaluate your best ideas, looking at the good and bad points and identifying what needs to be developed further
include user-group feedback to help you determine the best idea to develop further.

This stage is a 'progress check'. Draw together the design thinking of your initial ideas, gather third-party feedback and make a decision as to which is the most suitable design idea to develop.

What the exam board is looking for here is a formal conclusion to the initial ideas stage. You should present an objective evaluation of your design ideas, set against relevant specification points to determine their potential.

Your initial design ideas are 'raw' at this stage, and it is important to determine which ideas can be developed into workable solutions by testing them against your original specification criteria.

Your annotations should explain a lot of your design thinking, so you simply need to formalise this on a review sheet.

People's opinions

All design is concerned with people, and their opinions are useful in gaining another perspective on your ideas. Your designing should never occur 'in a vacuum' – we always need the views of other people, and their suggestions may influence further development of the product.

You could seek user-group feedback in a number of ways, including:

- questionnaires
- informal interviews
- 'sound bite' comments.

If you have identified a specific client for your product, you must record the client's opinions at this stage.

You should also consider sustainability issues when reviewing your best ideas. If you have used 'sustainability' as a heading in your original specification criteria, you can review your idea against this.

Example review work: formal review sheet

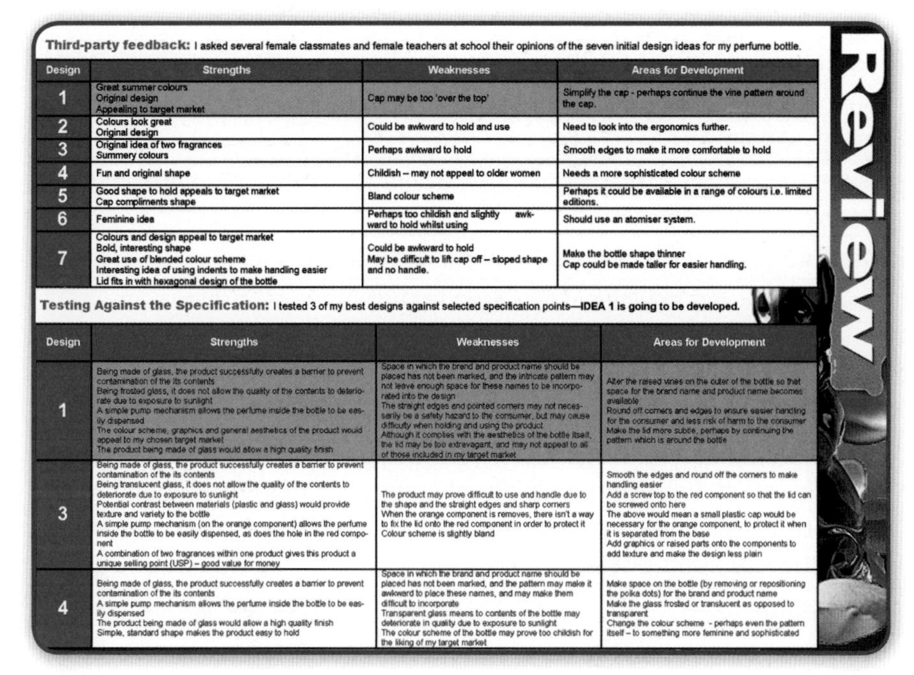

Third-party feedback: I asked several female classmates and female teachers at school their opinions of the seven initial design ideas for my perfume bottle.

Design	Strengths	Weaknesses	Areas for Development
1	Great summer colours Original design Appealing to target market	Cap may be too 'over the top'	Simplify the cap - perhaps continue the vine pattern around the cap.
2	Colours look great Original design	Could be awkward to hold and use	Need to look into the ergonomics further.
3	Original idea of two fragrances Summery colours	Perhaps awkward to hold	Smooth edges to make it more comfortable to hold
4	Fun and original shape	Childish – may not appeal to older women	Needs a more sophisticated colour scheme
5	Good shape to hold appeals to target market Cap compliments shape	Bland colour scheme	Perhaps it could be available in a range of colours i.e. limited editions.
6	Feminine idea	Perhaps too childish and slightly awk-ward to hold whilst using	Should use an atomiser system.
7	Colours and design appeal to target market Bold, interesting shape Great use of blended colour scheme Interesting idea of using indents to make handling easier Lid fits in with hexagonal design of the bottle	Could be awkward to hold May be difficult to lift cap off – sloped shape and no handle.	Make the bottle shape thinner Cap could be made taller for easier handling.

Testing Against the Specification: I tested 3 of my best designs against selected specification points—**IDEA 1 is going to be developed.**

Design	Strengths	Weaknesses	Areas for Development
1	Being made of glass, the product successfully creates a barrier to prevent contamination of the its contents Being frosted glass, it does not allow the quality of the contents to deterio-rate due to exposure to sunlight A simple pump mechanism allows the perfume inside the bottle to be eas-ily dispensed The colour scheme, graphics and general aesthetics of the product would appeal to my chosen target market The product being made of glass would allow a high quality finish	Space in which the brand and product name should be placed has not been marked, and the intricate pattern may not leave enough space for these names to be incorpo-rated into the design The straight edges and pointed corners may not neces-sarily be a safety hazard to the consumer, but may cause difficulty when holding and using the product Although it complies with the aesthetics of the bottle itself, the lid may be too extrevagant, and may not appeal to all of those included in my target market	Alter the raised vines on the outer of the bottle so that space for the brand name and product name becomes available Round off corners and edges to ensure easier handling for the consumer and less risk of harm to the consumer Make the lid more subtle, perhaps by continuing the pattern which is around the bottle
3	Being made of glass, the product successfully creates a barrier to prevent contamination of the its contents Being translucent glass, it does not allow the quality of the contents to deteriorate due to exposure to sunlight Potential contrast between materials (plastic and glass) would provide texture and variety to the bottle A simple pump mechanism (on the orange component) allows the perfume inside the bottle to be easily dispensed, as does the hole in the red compo-nent A combination of two fragrances within one product gives this product a unique selling point (USP) – good value for money	The product may prove difficult to use and handle due to the shape and the straight edges and sharp corners When the orange component is removes, there isn't a way to fix the lid onto the red component in order to protect it Colour scheme is slightly bland	Smooth the edges and round off the corners to make handling easier Add a screw top to the red component so that the lid can be screwed onto here The above would mean a small plastic cap would be necessary for the orange component, to protect it when it is separated from the base Add graphics or raised parts onto the components to add texture and make the design less plain
4	Being made of glass, the product successfully creates a barrier to prevent contamination of the its contents A simple pump mechanism allows the perfume inside the bottle to be eas-ily dispensed The product being made of glass would allow a high quality finish Simple, standard shape makes the product easy to hold	Space in which the brand and product name should be placed has not been marked, and the pattern may make it awkward to place these names, and may make them difficult to incorporate Transparent glass means to contents of the bottle may deteriorate in quality due to exposure to sunlight The colour scheme of the bottle may prove too childish for the liking of my target market	Make space on the bottle (by removing or repositioning the polka dots) for the brand and product name Make the glass frosted or translucent as opposed to transparent Change the colour scheme - perhaps even the pattern itself – to something more feminine and sophisticated

Moderator's comments

A formal review sheet is used as a 'progress check' to help decide which design to take forward to the development stage. Third-party feedback of all design ideas is recorded as 'sound bites' and three of the best ideas have been tested against relevant specification criteria. The learner has highlighted the design that they are taking forward.

Example review work: ongoing annotation

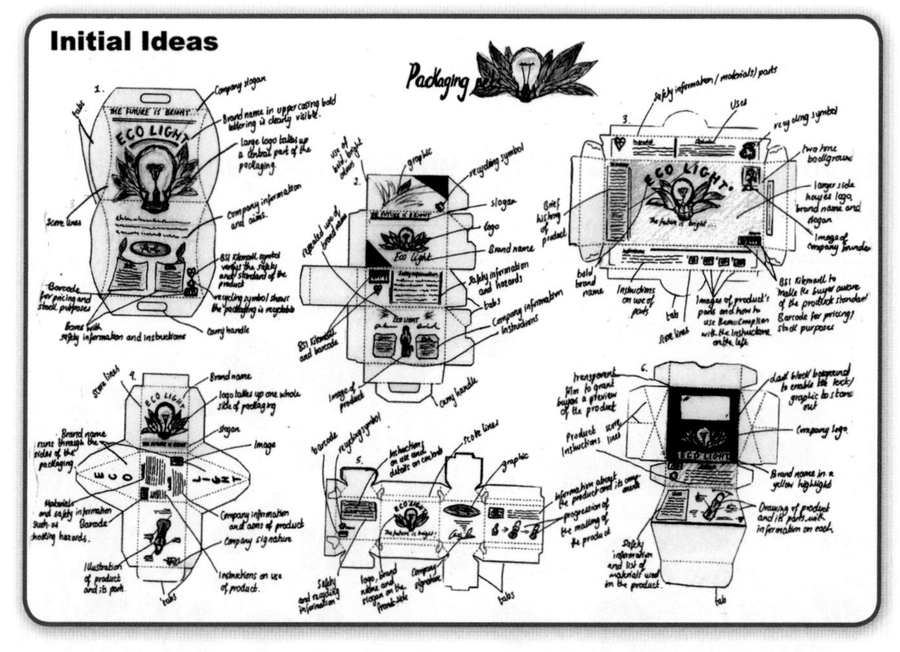

Moderator's comments

Ongoing annotation of initial design ideas will of course relate back to your original specification criteria. This learner has reviewed each idea as it has been drawn and has included third-party feedback. A formal review page will simply compile all of this information.

Objectives

- **Use** a range of communication techniques and media throughout the design, review and development stages.
- **Use** communication techniques with precision and accuracy.

2.3 Communication

To get top marks you will:
present your initial ideas using 2D and 3D sketches, with use of colour where appropriate
use ICT to design 2D graphic products where appropriate or to present the review of your ideas
use CAD to design 3D graphic products where appropriate
draw precise and accurate initial ideas.

Marks are awarded for communication techniques demonstrated throughout the design stages (initial ideas, review and develop only).

To be successful, you will present your design ideas in a logical format with clear sketches using a variety of communication techniques, such as 2D or 3D sketches, nets or exploded drawings, using shading and colour where appropriate.

Moderators want to be able to see the most important design decisions you have made, which could be highlighted areas on a busy design sheet.

There are many ways for you to present your initial design ideas, from thumbnail pencil sketches to computer-generated images. Don't be afraid to experiment.

The quality of your design work should enable clear communication of your design intentions.

Example communication work: pencil and marker sketches

Moderator's comments

This learner has used pencil and marker rendering which enables bold and effective communication of their design ideas. These ideas for brand identity are hand-drawn with precision and accuracy, and are fully annotated.

Example communication work: ICT

Masthead: Initial Designs

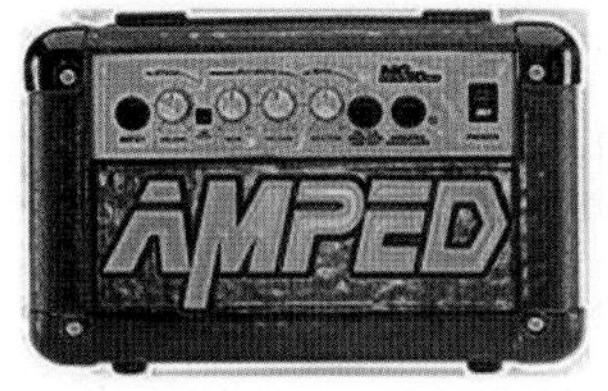

Design 1
This masthead design is based on a guitar amplifier; this is because all of the TMG will be fans of the genre of music that usually implements this tool, also some of the TMG will be musicians who use this piece of equipment, therefore they will instantly recognise this is a rock/metal magazine through the imagery created. The font and colour used for the word "Amped" is similar to a well known amplifier producer, this is because it again lets the TMG associate the masthead design with the type of magazine and the content

Design 2
This masthead has been designed around the image of a speaker booming out the name of the magazine, this is because the imagery created allows the TMG to associate the magazine with music. A black and white colour scheme has been used on the text, this is because it will create continuity on the front of the magazine also I believe it is an aesthetically appealing colour for my TMG.

Design 3
This masthead has been designed as a guitar arm; this is because it will be associated with the genre of the magazine by the intended TMG. The font of the text "Amped" has been produced in the style of a popular bands logo, this is because it will appeal to the TMG, and also they will know the magazine is a rock/metal magazine because of the genre of the music from the band.

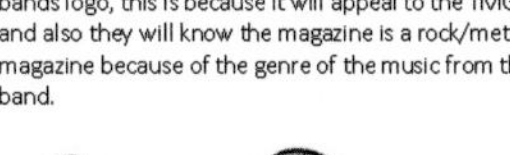

Design 4
This masthead has been designed to resemble a guitar lead; this is because it is a piece of equipment associated with music, so it will appeal to the TMG who are fans of music. The colours used are to ensure the masthead looks authentic like a guitar lead, this is again to make that association with music, as the magazines genre is rock/metal music.

Design 5
This masthead has been designed to look like a belt buckle; this is because it is a piece of clothing that is vaguely associated with rock/metal, so it will appeal to the TMG who are fans of this music genre. The colour gold is used to reflect luxury, as the magazine is to be of high quality, this is because this is what the TMG are expecting of the magazine as it will be fairly expensive. Parts of the belt have been drawn; this is to ensure the viewer can tell it is a belt buckle although when professionally produced, it will look more authentic.

Design 1
I have chosen design 1 for development, the main reason for this is because I believe the imagery created has the most association with music on first sight because an amplifier is a known instrument for music. Also I believe this masthead design has the greatest aesthetic appeal, as it is innovative and I have not seen a logo similar or the same as this design.

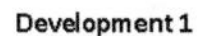

Development 1
For my first development I have simplified the design, this is because most magazine mastheads are very simple and aesthetically appealing. I have removed all of the text on the control panel of the amplifier; this is because it greatly simplifies the design whilst ensuring the viewer can still recognise the shape as an amplifier. I have also made the gold colour on the control panel solid; this again is because it adds to the simplicity of the design, intern making it look more like a magazine masthead.

Development 2
For my second development I have simplified the colour and texture in the design, this is because it will make my design look more like an amplifier. I have removed all of the different textures on the design this is because although the quality of the image has been decreased, the design has been dramatically simplified making it look more authentic like a masthead whilst retaining the amplifier shape. I have also changed the various colours on the design to a single orange; this again is because it makes the design look more like a masthead as mastheads usually consist of 1 or 2 colours.

Development 3
For my third and final development I have changed the colour in the design, this is because it will make my design appeal to more of my TMG. I have changed the orange originally selected to a blue, this is because I believe orange is more of a feminine colour and my TMG combines the 2 sexes. I believe that the blue I have selected is more rounded and will appeal to both male and female consumers; also I personally believe it looks more aesthetically appealing than orange.

Moderator's comments

ICT can be used to great effect to generate initial design ideas.This learner has used a graphics package to present a range of ideas for the masthead of a magazine they are designing. This graphics package has allowed the learner to experiment with typeface, visual effects and colour schemes in order to create a professional-looking image.

ResultsPlus
Watch out!

CAD can be part of the initial ideas stage but should not be over-used. The exam board is looking for the 'raw ideas' at this stage, and not a refined idea yet. If you are really confident in the use of CAD, and you can generate ideas quickly enough, then feel free to use it. However, CAD is generally better suited to the development stage.

Stage 3 Develop (15 marks)

Objectives

- **Develop** your initial design ideas into a final design proposal.
- **Use** scale modelling to test important aspects of your design idea as it progresses.
- **Evaluate** your ideas against relevant design criteria as they progress.

Controls

Develop

- You can produce rough sketches and gain user-group feedback when developing your ideas outside the classroom, and bring these into the class.
- Your developed ideas must be copied up into its final form in a classroom.
- You cannot make models at home – they must be made in a classroom/workshop.
- Your teacher can provide you with feedback to make sure that you have sufficiently developed your design idea using models and user group feedback.

ResultsPlus
Watch out!

You cannot access the high marks if you simply make minor and cosmetic changes to your original design idea. You must develop technical details.

3.1 Development

To get top marks you will:
improve your chosen design idea by developing technical details as well as cosmetic features
use scale models to refine and test your ideas as they progress, using block models and/or 3D software
continually review your designs against relevant specification criteria
include user-group/third-party feedback.

The most important part of the development stage is the refinement of one design idea in detail, and not simply with minor and cosmetic changes. The exam board is looking for you to develop your selected idea to a point where you produce a final design proposal for your product that addresses most of the points in your original specification criteria.

You should understand that 'to develop' means changing and modifying your ideas, which should include refining features of previous design ideas, in order to produce your final design proposal. Your final design proposal should be significantly different to, and should improve on, any previous initial design idea.

Design is about people. You must include user-group feedback in the development of your product so that it satisfies user requirements.

Modelling

Modelling plays a significant role in the design development cycle. Simple mock-ups or block models can be invaluable in determining whether a design is workable. For example, if you are designing a mobile phone, it would be very useful to produce rough Styrofoam models to get a feel for the product. It is sometimes difficult to visualise something in 3D just by looking at a 2D sketch. Vital information such as ergonomics may be obtained about a design if you can actually pick up a 3D model. However, there should always be a reason for modelling. For example to test ergonomics and proportions.

It is anticipated that the use of computer-aided design (CAD) will be far more relevant at this stage than at the initial ideas stage. If you use CAD to develop your idea, you may not need to produce a physical model at all. However, the testing of ergonomic features may require the user group to actually handle the product.

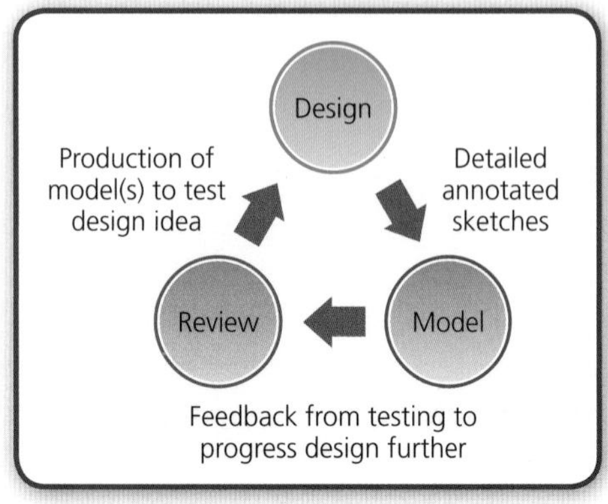

Figure 8.8: The design development cycle

Example development work: Styrofoam™ models

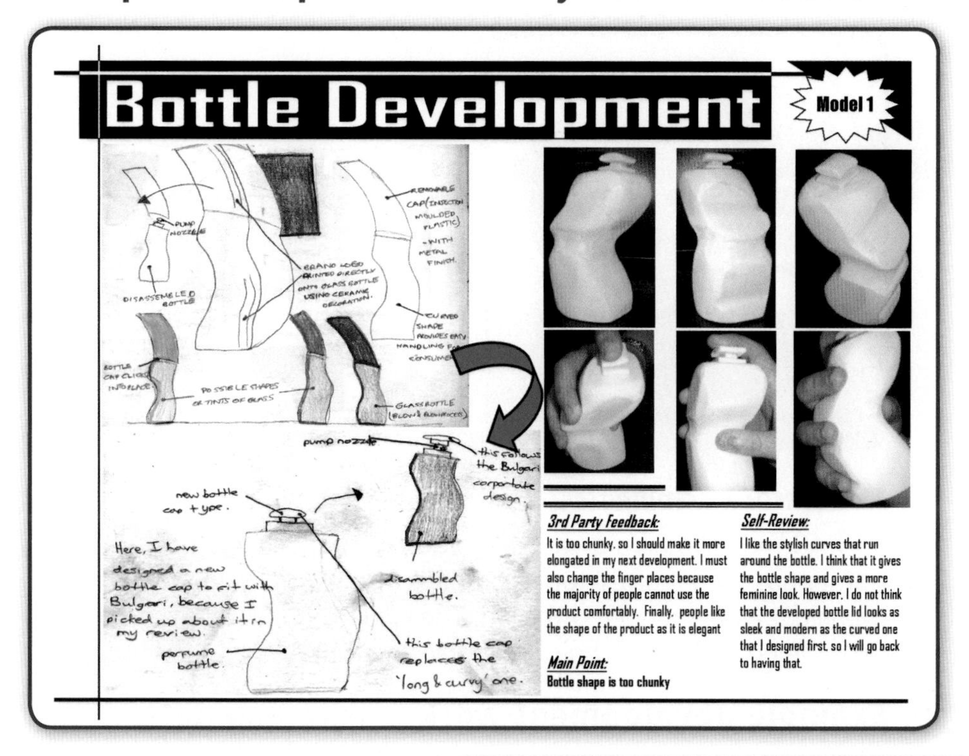

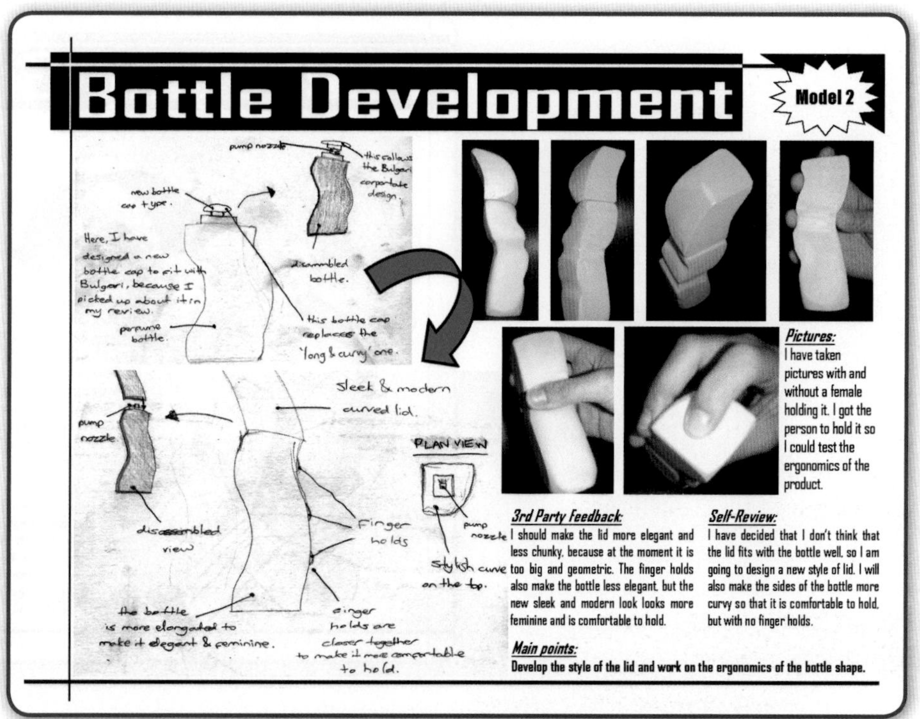

Moderator's comments

Development should include modelling to test your ideas as they progress. This learner has used a series of Styrofoam™ models to test the ergonomic features of a perfume bottle. Third-party feedback and self-review enables the learner to improve the design. The second model attempts to address problems that were highlighted by the first model.

Example development work: 3D CAD

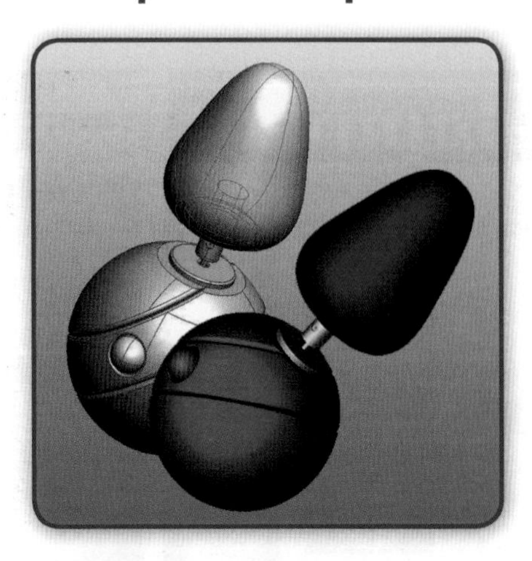

Moderator's comments

3D CAD modelling is a great way to develop an idea in detail. This learner has used ProDesktop to produce a solid model and then overlaid it with a rendered version. The result is the development of the perfume bottle both technically and aesthetically.

Example development work: packaging nets

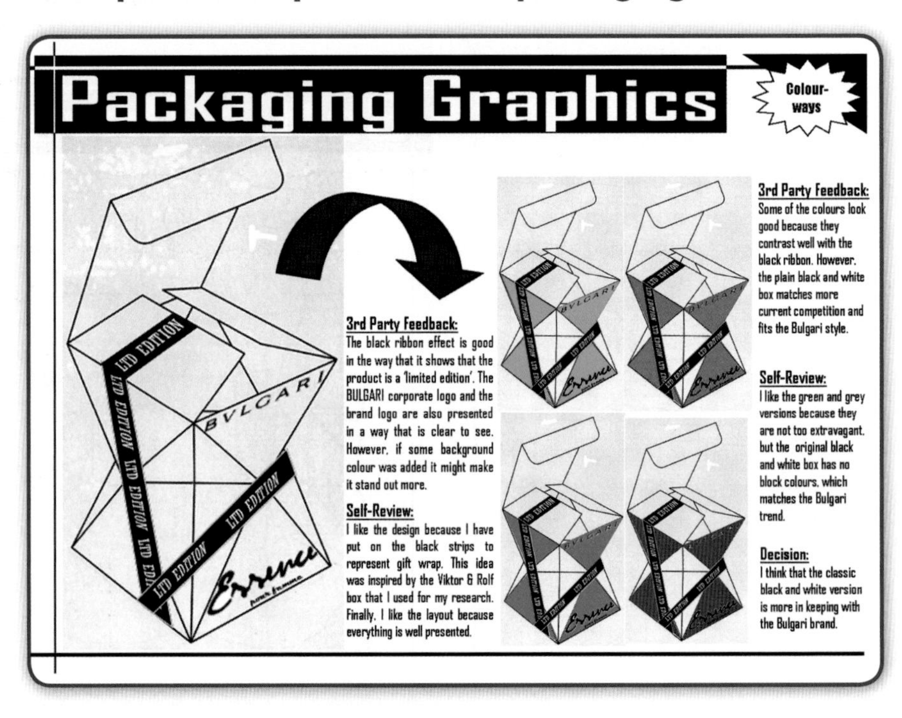

Moderator's comments

Making mock-ups of packaging nets is useful for checking dimensions and fit. A wide range of commercial packaging nets is available so you can develop an existing packaging net to best suit the product you have designed to fit inside it. This learner has clearly documented the modelling and testing of a commercial packaging net to contain a perfume bottle.

Example development work: colour

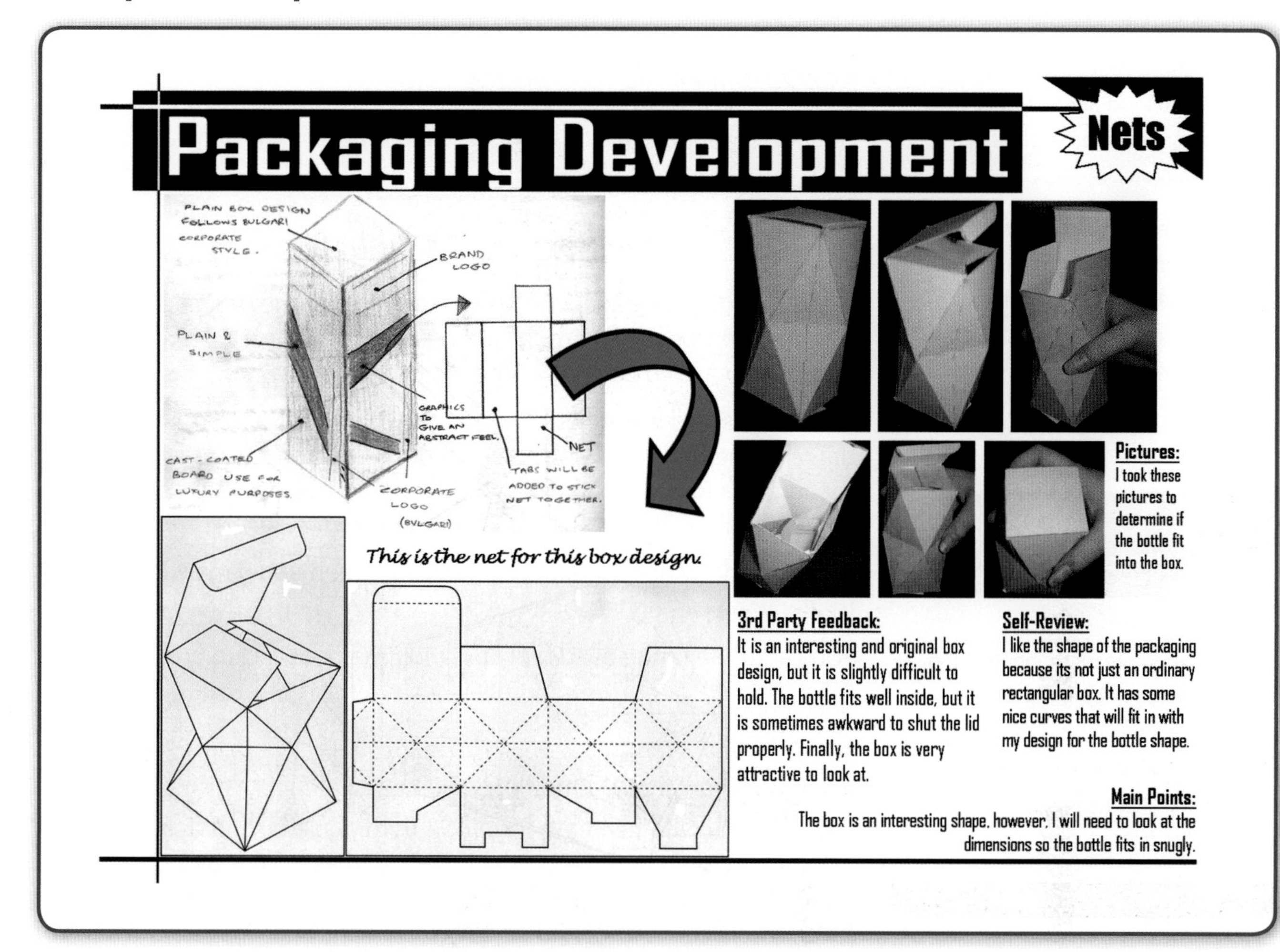

Moderator's comments

Exploring colourways can be important in graphic products, but it should not form the basis of your development; technical development is needed more. This learner has decided on an appropriate packaging net through a series of mock-ups, and can now experiment with colour as the technical development has finished.

Objectives

- **Present** your final design proposal in a format that communicates your design intentions.
- **Present** technical details of materials and/or components, processes and techniques relating to your final design proposal.

3.2 Final design

To get top marks you will:
clearly communicate your final design using a working drawing
include all information and major dimensions to enable a third party to make a model of the product
indicate the materials, components, processes and techniques that will be used to commercially manufacture the product.

In the final design stage, you draw the whole of the design activity together and clearly communicate the final, fully developed design proposal to a third party (somebody else).

The exam board is not necessarily looking for a set of formal technical drawings here. Marks will be awarded for the communication of all the information necessary for a third party to have a good chance of actually making the model of the product from it.

Your final drawings should be clearly annotated and dimensioned, so that anyone can understand them.

You can use a range of suitable drawing methods including:

- working drawings,
- exploded drawings
- pictorial drawings (isometric, planometric, etc.).

Information could be in the form of a third angle orthographic or isometric sketch of the product, with all major dimensions and components labelled and reference made to the materials and processes used. CAD would be extremely beneficial here as a detailed 3D sketch can easily be translated into a fully dimensioned working drawing.

ResultsPlus
Watch out!

A fully rendered presentation drawing of the final design is not required at this stage – it is the technical details that are most important. Don't waste time on a hand-drawn pictorial drawing, as there are no marks available for it.

Example final design work

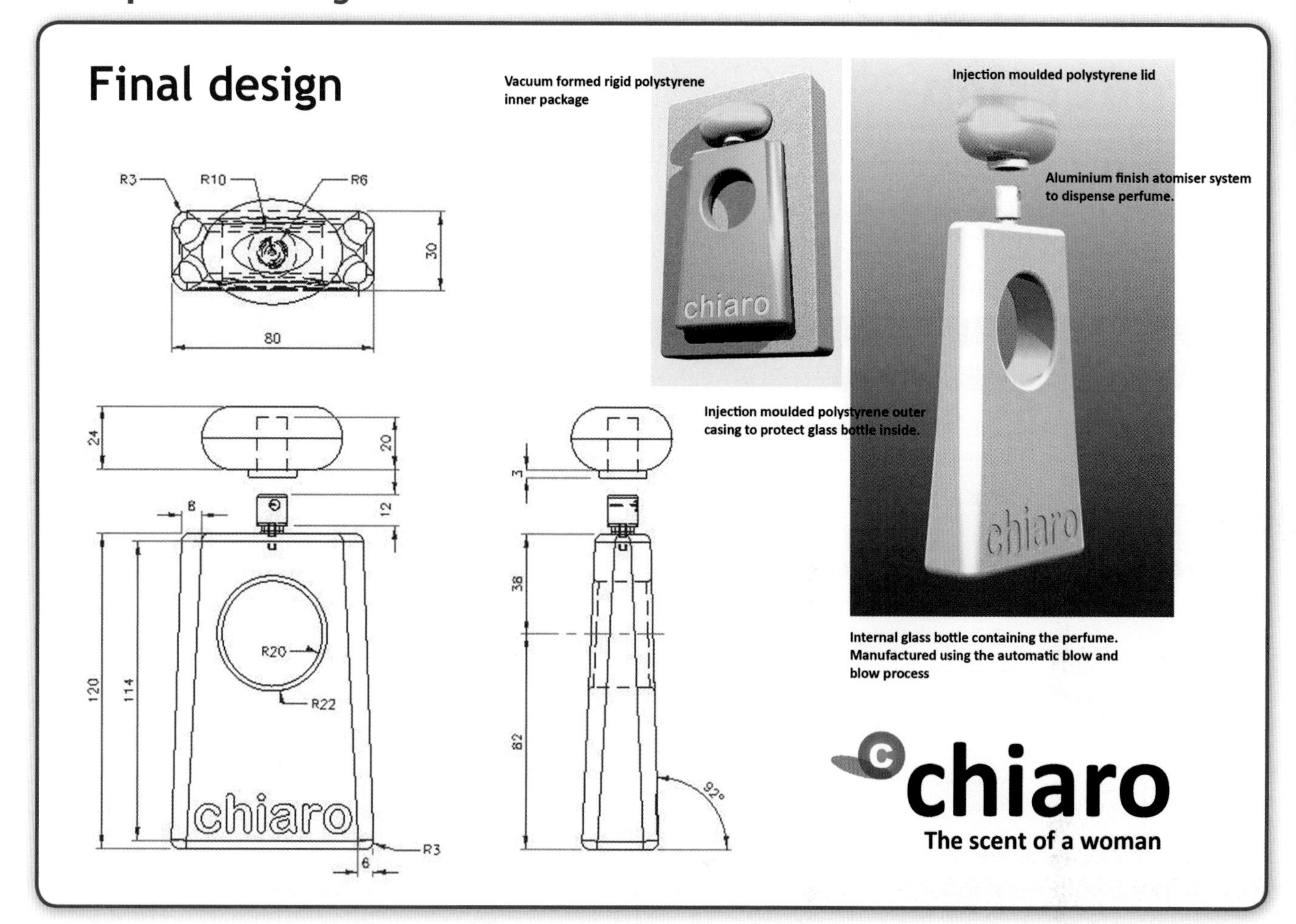

Moderator's comments

This learner's final design is presented on a single page, but it includes all the necessary drawings and information to get a clear idea of the product. A 3D CAD package has been used to construct the pictorial drawings and a fully dimensioned third angle orthographic drawing has been generated from this information. The working drawing contains enough technical details for a third party to actually make a model of the perfume bottle. The rendered pictorial drawings are clearly labelled, showing all of the materials and manufacturing processes that would be used to mass-produce the actual product.

Chapter 9 Make activity
Introduction

You will undertake a make activity covering the three stages and five assessment criteria listed in Table 9.1. The table also gives you an idea of the time you should allow for your work to meet each criterion, and of the number of pages you might produce for each.

Table 9.1: Outline of the make activity

Stages	Assessment criteria	Marks	Suggested times	Suggested number of pages
4. Plan	**4.1** Production plan	6	1–2 hours	1
5. Make	**5.1** Quality of manufacture	24	16 hours practical	1–2
	5.2 Quality of outcome	12		1
	5.3 Health and safety	2	Evidenced throughout make stage	N/A
6. Test and evaluate	**6.1** Testing and evaluation	6	1–2 hours	1–2
	TOTAL	**50**	**18–20 hours**	**4–6**

Please note that these are only suggested times and numbers of pages for each assessment criterion – they are not compulsory. However, we strongly recommend that you keep to the deadline of 20 hours for the whole of this make activity. The 16 hours of practical time does not include the writing-up of the manufacturing process – this should be seen as additional time. You should be able to achieve high marks for each assessment criterion within the suggested number of pages.

Starting points

If you have decided to produce a combined design and make project then you will now go on to plan, make, test and evaluate the product you designed in your design activity. However, if you are starting a different make activity at this stage, you will need one or more of the following:

- a working drawing provided by your teacher
- a photograph of the product, or the actual product, which will be copied or made to a suitable scale
- a manufacturing specification that includes relevant criteria to check performance and quality issues.

Remember, there are no design decisions to be made in this activity – it is your making skills that are being tested here. It is extremely important that you select a product to make that is sufficiently complex for you to access the high mark bands. For example, does your object require a wide range of skills and processes to be used in its manufacture? Your teacher will guide you through this selection process.

A further possibility is to make a product that you have designed from a completely different design task. The starting point for this make activity could be the working drawing of a product you designed from a previous, totally unrelated design task. For example, a learner has used perfume packaging for the design task in Year 11 but wants to make a point-of-sale display designed back in Year 10. This is fine as long as the point-of-sale display is sufficiently complex to give this learner access to the high mark bands.

Starting points: combined design and make activity

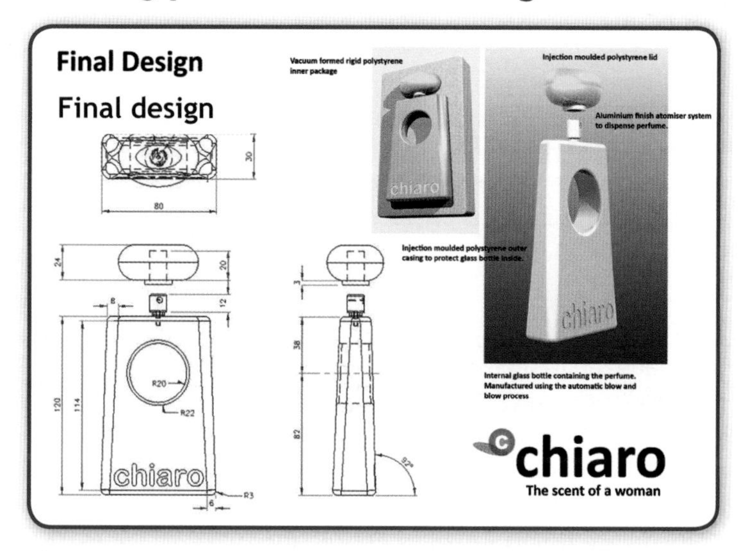

This example shows that a learner has decided to make a model of what they have created during their design activity. Their final design sheet shows a detailed working drawing for a perfume bottle that has several components and is suitably complex to access the high mark bands. They will use their original design specification to test and evaluate the final product model.

Starting points: separate make activity

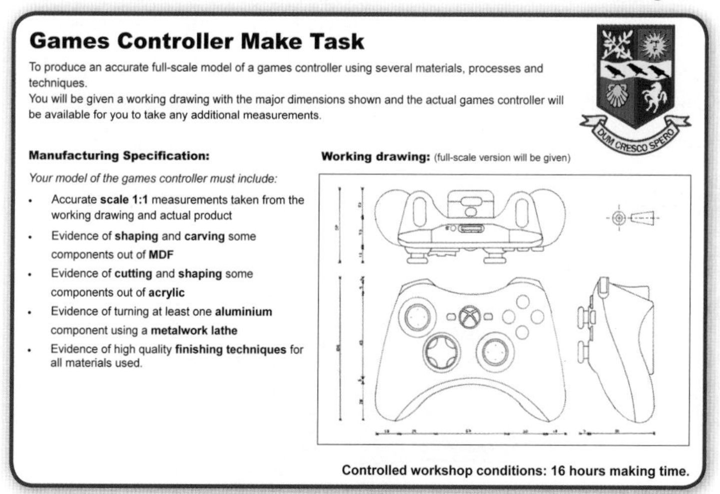

This example would fit nicely under the Edexcel set task heading 'Concept Design'. The teacher has given the learners a make task to model a games controller, which the majority of learners will be familiar with.

Here the teacher has provided learners with a working drawing with major dimensions shown, along with the opportunity to handle the actual product. Having access to the product means that learners can take additional measurements and use it to test ergonomic features such as curves while making their model. The manufacturing specification clearly indicates that learners need to use several materials, processes and techniques.

Controls

Make activity

Preparation

- You can undertake research and preparatory work outside the classroom without supervision. This does not form part of the controlled assessment time.

Write-up and making

- You must complete the making of your product under informal classroom supervision in accordance with health and safety regulations.
- Your teacher will be allowed to give demonstrations of all new processes and techniques.
- You are not allowed to take your practical work out of the classroom or workshop: for example, you will not be able to take practical work home.

Class Activity

1. List the making skills and processes that learners would be able to demonstrate if they followed this make task.
2. Which materials and processes would you use to make the following parts of the games controller:
   - main body?
   - analogue joysticks?
   - directional pad?
   - control buttons?
3. Read the assessment criteria for 'Quality of manufacture' (24 marks). Would a learner that produced the games controller be able to access the high mark band through this making task?

Stage 4 Plan (6 marks)

Objectives

- **Produce** a detailed production plan that considers the stages of manufacture for your product.
- **Describe** the stages during making where specific quality control procedures should take place.

Controls

Plan

- You can produce outline plans outside the classroom and bring them into the class.
- Your production plan must be copied up into best in a classroom.
- Your teacher can provide you with feedback to make sure that you have produced a sufficiently detailed production plan with specific quality control checks.

ResultsPlus
Watch out!

This should be the planning of the actual one-off product or concept model, and not cover how it would be made in volume using industrial and commercial processes.

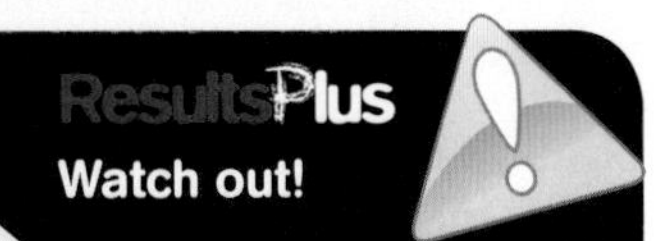

ResultsPlus
Watch out!

Your production plan must be forward-looking and not simply a diary of the making process completed after your model has been made.

4.1 Production plan

To get top marks you will:
produce a flowchart showing all the main stages in making your model in the correct order
indicate where specific quality control (QC) checks can take place
plan your making stages against time in order to meet deadlines.

The most important thing to have before you start planning is a detailed working drawing of the model you are attempting to make. This can be the product you designed in your design activity, or can be given to you by your teacher if you are doing separate design and make activities.

To ensure careful planning for the making of your one-off product, you must first produce a production plan showing the various stages of manufacture.

The production plan could take the form of a flowchart showing the stages of production in the correct order. Quality control (QC) points should also be clearly identified. The specific QC points should be named and described in detail, rather than simply using generic phrases such as 'check product for quality here'. For example, when cutting out materials, you should always check that you have cut out the part to the size required. This could be checked by measuring with a ruler. More precise measurements could be taken using a Vernier gauge.

Example production plan work: detailed working drawing

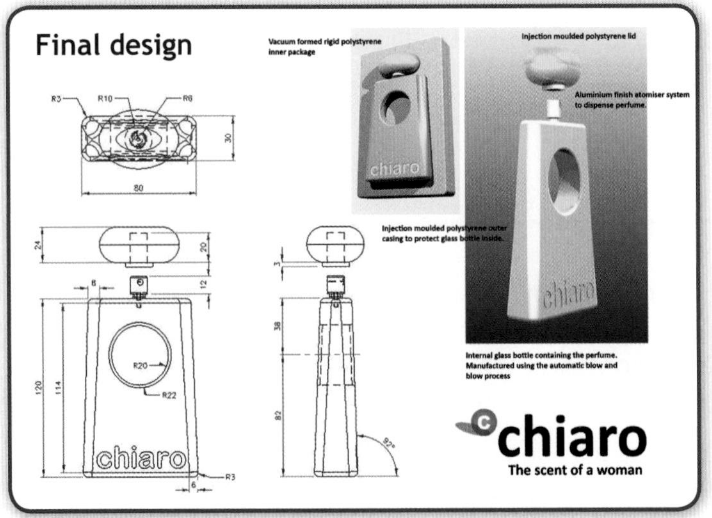

Moderator's comments

It is really important that you have a detailed working drawing of the product that you are going to make. From this drawing you will have to plan how you are actually going to make the model. Therefore all major dimensions and technical details have to be present.

Example production plan work: flowchart

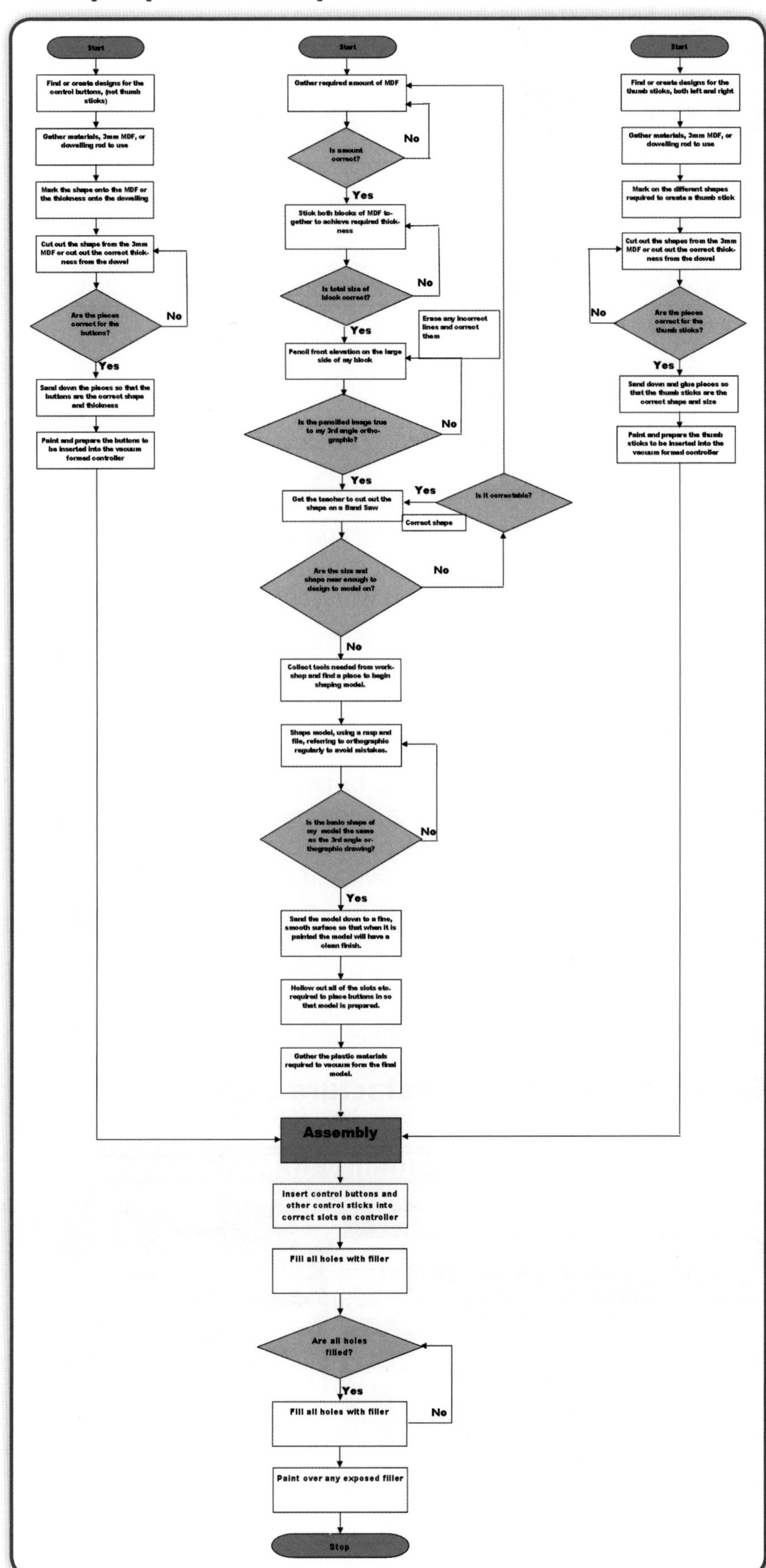

Moderator's comments

This learner's flowchart states the main stages in making the model in the correct sequence. Three different component parts feed into the assembly and finish of the overall product. Quality control checks are clearly indicated in order to ensure that a high-quality product can be made.

Time planning

In order to meet deadlines, you should plan your stages of manufacture against the time you have allocated. A simple way of showing this could be a Gantt chart like the one shown below. It is important that you use 'real-time' units such as hours, rather than days/weeks/lessons, which do not indicate measurable time.

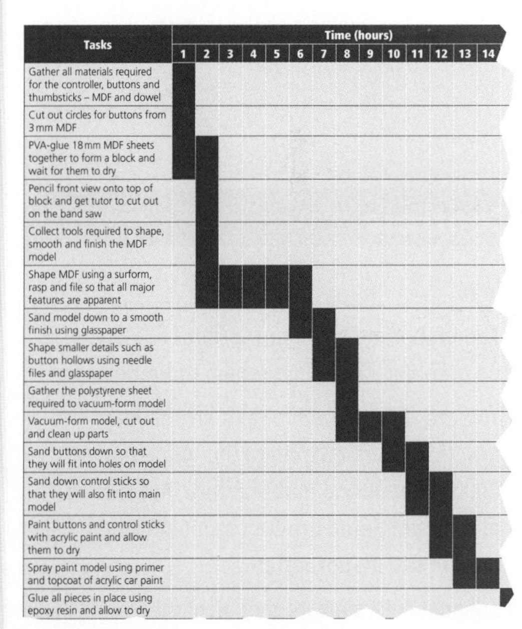

Moderator's comments

This learner's Gantt chart sets out the main tasks in making their model and allocates blocks of time in order to meet the deadline of 16 hours of practical time. A Gantt chart provides a visual reference that can be used to monitor your progress.

Stage 5 Make (38 marks)

Objectives

- **Attempt** a challenging making task involving the manufacture of several different components using a range of materials, equipment, techniques and processes.
- **Select** tools, equipment and processes, including CAD/CAM where appropriate, for specific uses.
- **Demonstrate** a detailed understanding of the working properties of materials you have selected for a specific use.
- **Demonstrate** a wide range of making skills with precision and accuracy.

Controls

Make

- The manufacture of your product(s) must be completed in a classroom or workshop
- You are not allowed to take any of your practical work outside of the workshop or classroom during manufacture, but testing of the completed product can be carried out off-site if appropriate.
- Edexcel must be sure that all of the practical work is your own, so you must take lots of photographs as you are making.
- Your tutor can give demonstrations to show you how to use specific tools, equipment and processes, but you must complete the actual work yourself.

5.1 Quality of manufacture

To get top marks you will:
select the correct tools, equipment and processes for the job with minimum guidance from your tutor
use a wide range of tools, equipment and processes
make several different component parts
demonstrate high-level making skills with component parts made with precision and accuracy
use computer-aided manufacture where appropriate.

This is an opportunity for you to be rewarded for your making skills.

First, you must ensure that your product provides you with an opportunity to make several different component parts using different materials and processes. Discuss this with your tutor.

You should then actually make your product to the best of your ability. The exam board is looking for your ability to select the right tools, equipment and processes for making a component part and the skills you demonstrate throughout the making process.

Finally, you should write up this stage by producing a full photographic record of the stages of manufacture, showing all the relevant processes in detail. This 'step by step' series of photographs should fully document all the tools, equipment and processes you used. Your photographs should be fully annotated to show all the decisions you made while making your product. For example, you should explain why you used a particular process in preference to another, and describe the problems you encountered and how they were overcome. Any descriptions you include should be brief and to the point.

Example quality of manufacture work

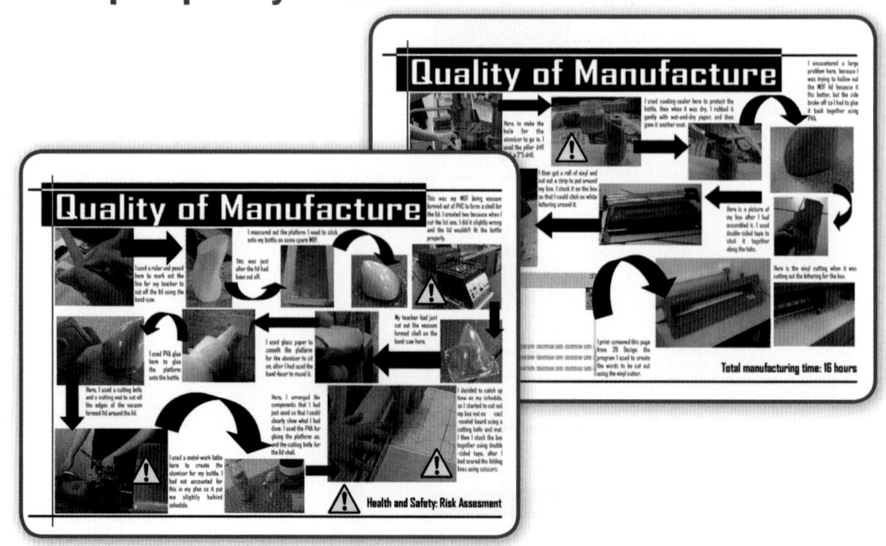

Moderator's comments

This learner has clearly documented the major stages in the manufacture of their 3D model of a perfume bottle and packaging net. It is presented in a logical order, and includes clear photographs and annotation explaining each stage, which makes it easy for a moderator to see just how much hard work has gone into making the product; which makes it easy for a moderator to see the levels of and range of skills used in making the product. Edexcel will not visit your school or college, so these pages are crucial in you showing a moderator the tools, equipment and processes you used and the quality of your manufacture.

ResultsPlus
Watch out!

You should not use more than 50 per cent CAM in your making activity so that other tools, equipment and processes can be fully evidenced.

Using a witness statement

Your teacher can support you further by using a witness statement. This provides evidence of their observation of your practical work throughout the making process. The witness statement will:

- state the tools, equipment and processes that you have demonstrated
- show the amount of guidance you were given when selecting tools, equipment and processes
- determine the skill you demonstrated in their use.

This will enable your teacher to support you, as they know you better than a moderator does. A moderator will always try to 'agree' the marks awarded if these are fully supported and justified.

The writing-up of this stage will not form part of the time allocated for controlled assessment. Making time will be devoted entirely to the actual making of your product, not its documentation.

Example witness statement

Component part	Tools, equipment and processes demonstrated	Selection			Skill		
		With guidance	Some guidance	Independently	Little attention to detail	Attention to detail	Precision and accuracy
Hardwood main body of radio	Turning on wood lathe			X		X	
Aluminium ON/OFF buttons	Turning and knurling using centre lathe		X			X	
Acrylic stand	Cutting using Hegner saw, filing and thermoforming in oven			X			X

Moderator's comments

The teacher has clearly named each component part and outlined which tools, equipment and processes the learner has used during the making process. A simple 'X' indicates the amount of guidance given to the learner and the level of skills demonstrated when making each component part.

Objectives

- **Produce** high-quality components that are accurately assembled and well finished to produce a high-quality product overall.
- **Produce** a completed product that is 'fully functional' as a graphic product.

ResultsPlus
Watch out!

'Fully functional' in terms of a graphic product usually means a concept model that fully communicates your design intention to a third party: for example, an architectural model made to a suitable scale and accurate in detail. It does not have to be the real thing!

5.2 Quality of outcome

To get top marks you will:
make each component part to a very high quality
ensure that each component part is well finished
assemble all your components parts together accurately
produce a product model that communicates your design intentions to a third party.

It is important to remember here that the making of your product is worth 36 marks in total, but making is divided into two different parts. During 'quality of manufacture' (worth 24 marks) it was your making skills that were being assessed, but in this part (worth 12 marks) it is the actual quality of your final outcome that is being assessed. This will include the manufacture of all the different component parts, assembling all these parts together and your final finished product.

Your final assembled and finished product should be fit for purpose and should meet the requirements of your original design specification or working drawings and manufacturing specification. If you ran out of time during the making process, it is the quality of the manufacture of individual components that will gain you some marks.

Although your project folder may be sent off to Edexcel for moderation, your actual product will not. You must be sure to take several photographs of the final outcome, showing its various details and different views, so that the moderator gets a greater feel for the quality of your product model.

Example quality of outcome work: 3D model

Moderator's comments

This learner's concept design model of a speaker uses a range of materials and processes. The pyramid shape is made from MDF with card speaker cones, while the base has been carefully shaped from Styrofoam. The pyramid and base are joined together using a threaded aluminium bar. The end result is an accurate and well-finished model.

Example quality of outcome work: CAM

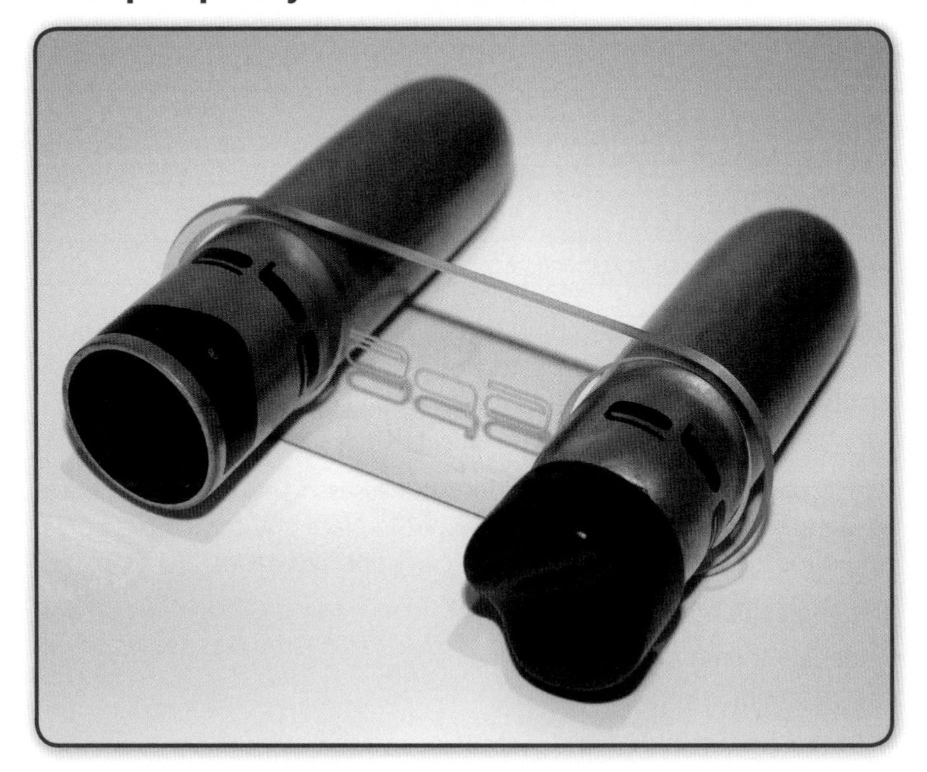

Moderator's comments

This learner's packaging design model of 'his and hers' perfume and aftershave bottles includes the use of computer-aided manufactured component parts. The acrylic stand has been laser-cut and etched, and the brand identity has been applied to the surface of the bottle shapes using vinyl-cut stickers. The learner has also used a wood lathe to turn the bottle shapes.

Remember, you should not use more than 50 per cent CAM in your making activity so that you can demonstrate your skills in using other tools, equipment and processes.

Example quality of outcome work: 2D model

Moderator's comments

This student's poster is part of a community awareness campaign for a London Borough. Do not underestimate the wide range of skills and processes that you need to produce a high-quality 2D product. Here, the student has taken an original digital photograph and used a professional software package to manipulate the image and create other graphics in order to produce a professional looking piece of work.

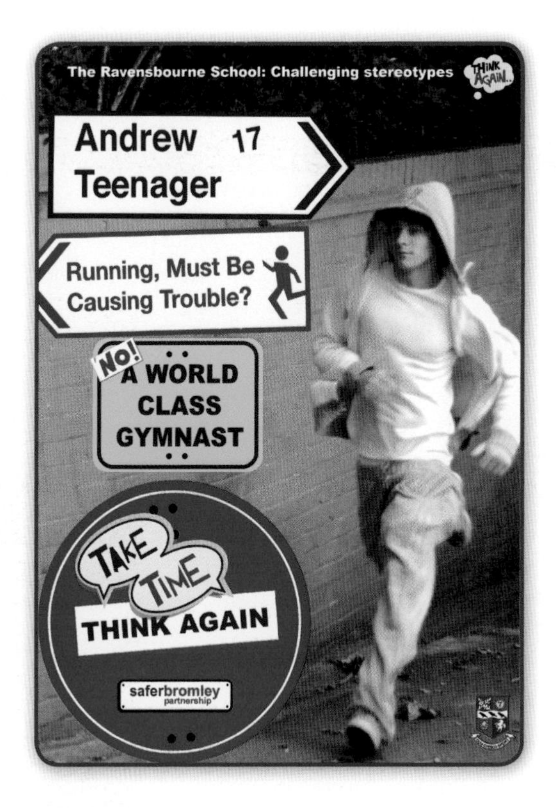

Objectives

- **Demonstrate** a high level of safety awareness when making your product.

ResultsPlus
Watch out!

Your tutor will award the marks for health and safety based on their observations of you during the making process. You do not need to record risk assessments for your make activities.

5.3 Health and safety

To get top marks you will:
take into account the risk assessments of relevant tools, equipment and machinery before using them
follow tutor instruction carefully when using tools, equipment and machinery
use tools, equipment and machinery with great care and attention.

During the making process, you should always be thinking about your own personal safety and the safety of others around you. No one wants accidents to happen and most can be avoided by assessing the risks before you start any making activities.

Your school or college must carry out risk assessments of their facilities to identify any potential hazards and put in place any control measures to reduce the risk of injury. They will carry out five stages:

1 Identify the hazard.
2 Identify the people at risk.
3 Evaluate the risks.
4 Decide on suitable control measures.
5 Record risk assessment.

Hazard	Risk	Control measure
Potential (of risk) from a substance, machine or operation	**Reality** (of harm from the hazard)	**Action** taken to minimise the risks to people

Your teacher will make you aware of any potential hazards and give you instruction in using any potentially dangerous tools, equipment and machinery. Listen carefully to these instructions and always act in a safe and responsible manner.

These marks are to be awarded by the teacher who is supervising your controlled assessment. No formal records are required, such as risk assessments of equipment. However, it is a good idea for you to annotate your photographs suitably to make reference to health and safety procedures, to remind your tutor of your care and attention to health and safety issues.

Example health and safety work

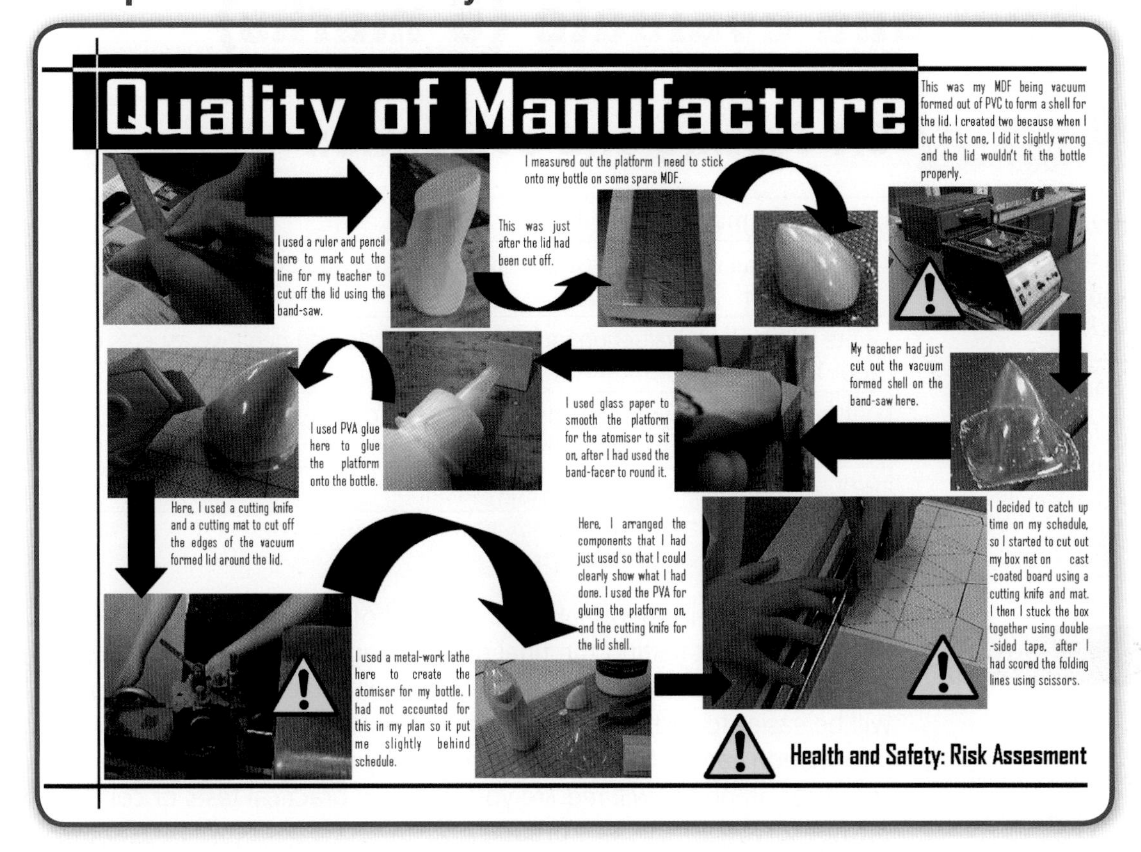

Moderator's comments

During the making process, photographic evidence can demonstrate that you have followed the correct health and safety procedures. Although you don't need to produce formal risk assessments, it is probably a good idea to flag up stages where you have really taken into account health and safety issues. For example, this learner has included 'warning triangles' to indicate where they have had to be particularly careful and follow the necessary risk assessments for that piece of equipment (vacuum former) and machinery (metal lathe).

Stage 6 Test and evaluate (6 marks)

Objectives

- **Devise** and carry out a range of suitable tests to check the performance and quality of the final product.
- **Evaluate** your final product objectively with reference to specification points, user-group feedback and sustainability issues.

Controls

Test and evaluate

- You can gain user-group feedback on your finished product outside the classroom if this is appropriate when testing against specifiication points and bring your findings into the class.
- The write-up of your testing and evaluation must be completed in a classroom.
- Your teacher can provide you with feedback to make sure that you have sufficiently tested and evaluated your final product.

ResultsPlus
Watch out!

If you are undertaking combined design and make activities, you should use your original design specification to test against. If you are undertaking separate design and make activities, you should use your manufacturing specification to test against.

6.1 Testing and evaluation

To get top marks you will:
test your final model against your design specification or manufacturing specification
write a balanced evaluation include both positive and negative aspects of your final model
get feedback from your user group regarding quality and performance issues
include sustainability issues in your evaluation.

Testing

Once you have finished making the model of your product, you will be ready to start testing the final outcome against your original design specification or manufacturing specification. Some specification points will be easier to test than others: for example, where quantitative information can be tested, such as checking dimensions. Other specification points will require you to devise practical tests or conduct further surveys of the user group.

Test questions

Use one of your specification points to create a question.

Specification point: *The box must fulfill the general packaging requirements by containing, advertising and protecting the bottle.*

Question: *Did my package contain, advertise and protect the perfume bottle?*

To answer this you need to study your product in detail. For example, was the packaging net strong enough to contain and protect the product? This could be to do with quality or performance.

Specification point: *The product must have a strong brand identity and incorporate the brand name 'Duo'.*

Question: *How did I create a strong brand identity incorporating the brand name 'Duo'?*

Here you will have to determine how to qualify your results. For example, you must *show* why it had a strong brand identity perhaps by asking other people or 'third-party' user testing. You could survey a sample from the user or target group identified in your specification about the effectiveness of the brand identity of the product, to gain an impression of the market's reaction.

Certain specification points will include quantitative information that can be tested easily: for example 'The board game must be no larger than 600mm × 600mm', which can be measured.

Example testing work

Testing Against the Specification

Perfume Bottle: I am very pleased with the final outcome for the perfume bottle. Although its an MDF model, I have designed it to be commercially produced so the design fits industrial processes such as the automatic blow and blow process for mass producing glass bottles and injection moulding of polymer components.

Specification Criteria		Have they been met?
Form	Incorporate the corporate name 'Ralph Lauren' as well as a suitable brand name to represent the product. Be attractive to my target market of women by having a sexy/ sophisticated style and colour scheme to ensure that they can easily connect with the new product. Be unique so that it competes successfully with other fragrances that are currently available on the market.	The bottle clearly displays the Ralph Lauren logo along with the brand name of the fragrance. The colour scheme has been designed to appeal to women using pastel colours to look feminine. There are no other pyramid shaped perfume bottles on the market at this time so I think it's unique.
Function	Be able to contain 80ml of liquid perfume in an inert glass bottle that acts as a barrier to contamination so that the fragrance remains fresh. Fulfil packaging requirements by presenting the perfume well aesthetically for marketing purposes i.e. crystal clear glass packaging. Easily dispense measured amounts of fragrance using an atomiser system. Protect the contents by using a sufficient thickness of glass which will not shatter easily if dropped.	Although my model is made from MDF, the actual bottle would be made from glass so it would keep the fragrance fresh. The atomiser system would be sourced from a separate company and would undergo quality control procedures to test whether it consistently dispenses the same amount of fragrance each time.
User Requirements	Be comfortable to hold in the user's hand (ergonomics) and easily dispense the fragrance without stretching fingers. Be freestanding so that it can be kept on a dressing table or in a bathroom cabinet. Incorporate a cap to cover the atomiser system to keep the nozzle clean and clear from dust and dirt.	Although it looks as if a pyramid shape would be uncomfortable to hold, it is actually okay to handle due to its rounded corners. The square base of the pyramid makes it very stable. I have incorporated a cap to keep the nozzle clean.
Performance Requirements	Be subject to quality tests and inspections using quality control systems to ensure that the consumer is receiving a product at the highest possible quality. Have a water–tight seal between the neck of the atomiser system and bottle to prevent leakage. Dispense a measured amount of fragrance with each press.	All components would be subject to rigorous QC tests and checks to assure the quality of the final product. Components coming from other manufacturers would still have to fulfil exact tolerances as laid down by Ralph Lauren or orders would be cancelled.
Material and Component Requirements	Be made to a high quality from suitable materials and components that are durable and give an aesthetically pleasing finish.	Both glass and polystyrene are durable. The colouring to the glass bottle and cap give an aesthetically pleasing effect.
Scale of Production and Cost	Be mass produced to satisfy international markets that Ralph Lauren currently competes in. Be manufactured using industrial processes such as automatic blow and blow for the glass bottle and injection moulded components such as the cap. Retail at around £35 to fit into the current Ralph Lauren range of fragrances.	The bottle is capable of being mass produced using the automatic blow and blow process for the glass bottle and injection moulding for the polystyrene cap. I believe that these processes would keep manufacturing costs down therefore, making profit for Ralph Lauren at a retail price of £35.
Sustainability	Be easily recyclable by separating the glass bottle from the atomiser and cap components. Use recycled plastic for components such as the cap which can be given a surface finish to disguise the grey look i.e. chrome effect.	The atomiser may not be easily removed as it has to be a water-tight seal so it may not be easily recycled. The cap is made from recycled polystyrene and coloured for visual effect.

Moderator's comments

This learner has followed combined design and make activities and therefore has an original design specification to test against. If you have made something different to what you designed, you should have been given (or devised yourself) a manufacturing specification to test against.

Evaluation

Evaluation is an ongoing process, which should be evident throughout your design folder. The final evaluation should consider all these comments and offer an honest opinion of the final product. Quality and performance issues are important at this stage and will need to be discussed. No design is perfect, so do not presume that yours is! Be honest and describe what you do not like about your design. This is intended to identify the weak points to address if the design were to be taken further. There will be also positive aspects of your project that you will need to highlight. Explain things that went particularly well in detail to illustrate your knowledge, skills and understanding. To access the high mark band, you must include relevant sustainability issues as in the life cycle assessment in Figure 9.1.

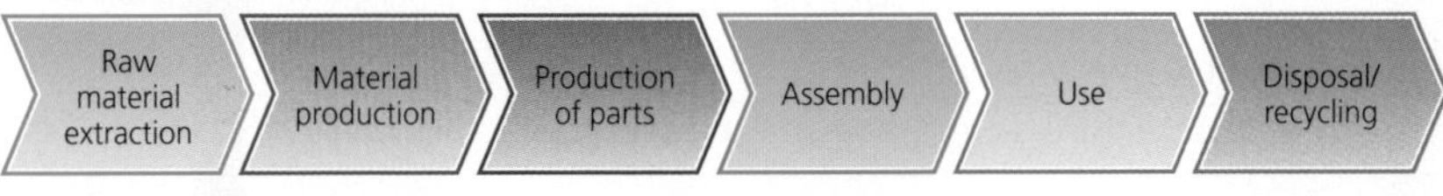

Figure 9.1: The stages in a life cycle assessment

Example evaluation work

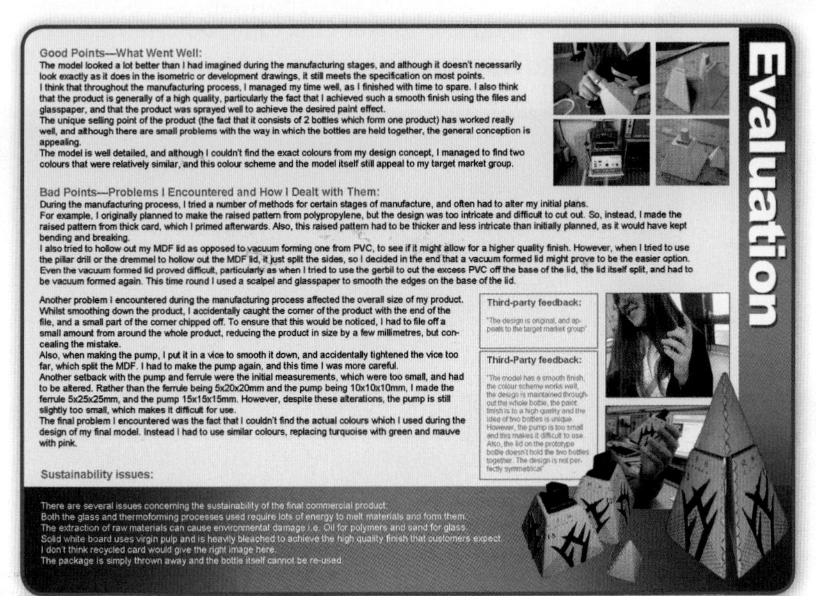

Evaluation

Good Points—What Went Well:

The model looked a lot better than I had imagined during the manufacturing stages, and although it doesn't necessarily look exactly as it does in the isometric or development drawings, it still meets the specification on most points.

I think that throughout the manufacturing process, I managed my time well, as I finished with time to spare. I also think that the product is generally of a high quality, particularly the fact that I achieved such a smooth finish using the files and glasspaper, and that the product was sprayed well to achieve the desired paint effect.

The unique selling point of the product (the fact that it consists of 2 bottles which form one product) has worked really well, and although there are small problems with the way in which the bottles are held together, the general conception is appealing.

The model is well detailed, and although I couldn't find the exact colours from my design concept, I managed to find two colours that were relatively similar, and this colour scheme and the model itself still appeal to my target market group.

Bad Points—Problems I Encountered and How I Dealt with Them:

During the manufacturing process, I tried a number of methods for certain stages of manufacture, and often had to alter my initial plans.

For example, I originally planned to make the raised pattern from polypropylene, but the design was too intricate and difficult to cut out. So, instead, I made the raised pattern from thick card, which I primed afterwards. Also, this raised pattern had to be thicker and less intricate than initially planned, as it would have kept bending and breaking.

I also tried to hollow out my MDF lid as opposed to vacuum forming one from PVC, to see if it might allow for a higher quality finish. However, when I tried to use the pillar drill or the dremmel to hollow out the MDF lid, it just split the sides, so I decided in the end that a vacuum formed lid might prove to be the easier option. Even the vacuum formed lid proved difficult, particularly as when I tried to use the gerbil to cut the excess PVC off the base of the lid, the lid itself split, and had to be vacuum formed again. This time round I used a scalpel and glasspaper to smooth the edges on the base of the lid.

Another problem I encountered during the manufacturing process affected the overall size of my product. Whilst smoothing down the product, I accidentally caught the corner of the product with the end of the file, and a small part of the corner chipped off. To ensure that this would be noticed, I had to file off a small amount from around the whole product, reducing the product in size by a few millimetres, but concealing the mistake.

Also, when making the pump, I put it in a vice to smooth it down, and accidentally tightened the vice too far, which split the MDF. I had to make the pump again, and this time I was more careful.

Another setback with the pump and ferrule were the initial measurements, which were too small, and had to be altered. Rather than the ferrule being 5x20x20mm and the pump being 10x10x10mm, I made the ferrule 5x25x25mm, and the pump 15x15x15mm. However, despite these alterations, the pump is still slightly too small, which makes it difficult for use.

The final problem I encountered was the fact that I couldn't find the actual colours which I used during the design of my final model. Instead I had to use similar colours, replacing turquoise with green and mauve with pink.

Third-party feedback:

"The design is original, and appeals to the target market group"

Third-Party feedback:

"The model has a smooth finish, the colour scheme works well, the design is maintained throughout the whole bottle, the paint finish is to a high quality and the idea of two bottles is unique. However, the pump is too small and this makes it difficult to use. Also, the lid on the prototype bottle doesn't hold the two bottles together. The design is not perfectly symmetrical"

Sustainability issues:

There are several issues concerning the sustainability of the final commercial product:

Both the glass and thermoforming processes used require lots of energy to melt materials and form them.

The extraction of raw materials can cause environmental damage i.e. Oil for polymers and sand for glass.

Solid white board uses virgin pulp and is heavily bleached to achieve the high quality finish that customers expect.

I don't think recycled card would give the right image here.

The package is simply thrown away and the bottle itself cannot be re-used.

Moderator's comments

This learner has also presented a well-balanced (objective) evaluation by discussing both the positive and negative aspects of the model. To access the high marks, the learner has also included sustainability issues and third-party feedback.

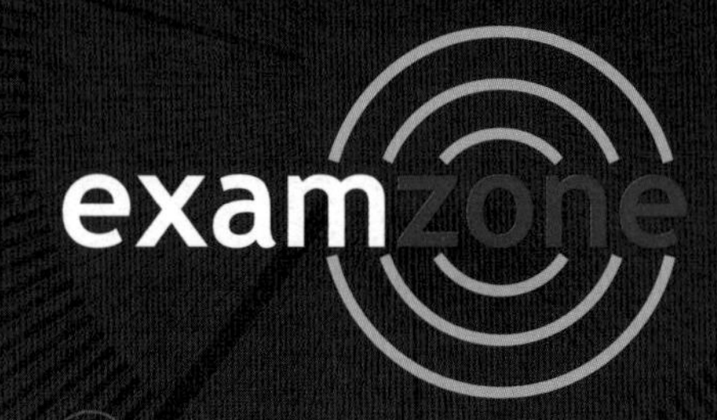

Welcome to ExamZone! Revising for your exams can be a daunting prospect. In this section of the book we'll take you through the best way of revising for your exams, step by step, to ensure you get the best results that you can achieve.

Zone In!

Have you ever become so absorbed in a task that it suddenly feels entirely natural? This is a feeling familiar to many athletes and performers: it's a feeling of being 'in the zone' that helps you focus and achieve your best. Here are our top tips for getting in the zone with your revision.

UNDERSTAND IT

Understand the exam process and what revision you need to do. This will give you confidence but also help you to put things into proportion. These pages are a good place to find some starting pointers for performing well at exams.

DEAL WITH DISTRACTIONS

Think about the issues in your life that may interfere with revision. Write them all down. Think about how you can deal with each so they don't affect your revision. For example, revise in a room without a television, but plan breaks in your revision so that you can watch your favourite programmes. Be really honest with yourself about this – lots of learners confuse time spent in their room with time revising. It's not at all the same thing if you've taken a look at Facebook every few minutes or taken mini-breaks to send that vital text message.

GET ORGANISED

If your notes, papers and books are in a mess you will find it difficult to start your revision. It is well worth spending a day organising your file notes with section dividers and ensuring that everything is in the right place. When you have a neat set of papers, turn your attention to organising your revision location. If this is your bedroom, make sure that you have a clean and organised area to revise in.

BUILD CONFIDENCE

Use your revision time not just to revise the information you need to know, but also to practise the skills you need for the examination. Try answering questions in timed conditions so that you're more prepared for writing answers in the exam. The more prepared you are, the more confident you will feel on exam day.

FRIENDS AND FAMILY

Make sure that they know when you want to revise, and even share your revision plan with them. Help them to understand that you must not get distracted. Set aside quality time with them, when you aren't revising or worrying about what you should be doing.

KEEP HEALTHY

During revision and exam time, make sure you eat well and exercise, and get enough sleep. If your body is not in the right state, your mind won't be either – and staying up late to cram the night before the exam is likely to leave you too tired to do your best.

The key to success in exams and revision often lies in the right planning. Knowing what you need to do and when you need to do it is your best path to a stress-free experience. Here are some top tips in creating a great personal revision plan.

My plan

1. Know when your exam is
Find out your exam dates. You should have been given an exam timetable but if not go to www.edexcel.com/iwantto/pages/dates.aspx to find all final exam dates, and check with your teacher. This will enable you to start planning your revision with the end date in mind.

2. Know your strengths and weaknesses
At the end of each chapter complete the 'You should know' checklist. Highlight the areas that you feel less confident on and allocate extra time to spend revising them.

3. Personalise your revision
This will help you to plan your personal revision effectively by putting a little more time into your weaker areas. Use your trial exam results and any further tests that you have taken as a self-assessment check on your progress to date

4. Set your goals
Once you know your areas of strength and weaknesses you will be ready to set your daily and weekly goals.

5. Divide up your time and plan ahead
Draw up a calendar, or list all the dates, from when you start your revision through to your exams.

6. Know what you're doing
Break your revision down into smaller sections. This will make it more manageable and less daunting. You might do this by using the Edexcel GCSE Graphic Products specification, or the chapter objectives, or headings within the chapter.

7. Link it together
Also make time for considering how Chapters interrelate.
For example, when you are revising Chapter 6 Sustainability 'minimising waste production' it would be sensible to cross reference the '4 Rs' to the materials listed in Chapter 1Materials and components. You could consider the impact on the environment of the extraction, production and use of all of the materials listed.

8. Break it up
Revise one small section at a time, but ensure that you give more time to topics that you have identified weaknesses in.

9. Be realistic
Be realistic in how much time you can devote to your revision, but also make sure you put in enough time. Give yourself regular breaks or different activities to give your life some variety. Revision need not be a prison sentence!

10. Check your progress
Make sure you allow time for assessing progress against your initial self-assessment. Measuring progress will allow you to see and celebrate your improvement, and these little victories will build your confidence for the final exam.

Finally – stick to your plan!

Know Zone
Chapter 1 Materials and components

Topic 1.1: Paper and board

Recall:

There are two types of paper that you need to know about:

1. Cartridge paper
2. Tracing paper

There are four types of board that you need to know about:

1. Folding boxboard
2. Corrugated board
3. Solid white board
4. Foil-lined board

And finally . . . . a packaging laminate consisting of:

- Polyethylene
- Aluminium foil
- Paper board

Explain:

You need to know the properties, advantages/disadvantages and uses of the paper and board listed above. Properties can include aesthetic, functional and mechanical associated with each material. Try to organise your notes under these headings.

Apply:

You should apply your knowledge and understanding of paper and boards to specific uses. For example, explain why solid white board is used to package designer perfumes.

Examiner's tip

Many of the disadvantages of paper and board involve environmental issues in its manufacture and disposal.

Topic 1.2: Metals

Recall:

There are three types of metals that you need to know about:

1. Steel – ferrous metal
2. Aluminium – non-ferrous metal
3. Tin – non-ferrous metal

Explain:

You need to know the properties, advantages/disadvantages and uses of the metals listed above. Properties can include aesthetic, functional and mechanical associated with each material. Try to organise your notes under these headings.

Apply:

You should apply your knowledge and understanding of metals to specific uses. For example, explain why aluminium is used for road signs instead of steel.

Examiner's tip

Metals are mostly used in graphic products for packaging and signage.

Topic 1.3: Polymers

Recall:

There are six types of polymer that you need to know about:

1. Acrylic
2. PET
3. PVC
4. Polypropylene
5. Polystyrene (rigid and expanded types)
6. Styrofoam

Explain:

You need to know the properties, advantages/disadvantages and uses of the polymers listed above. Properties can include aesthetic, functional and mechanical associated with each material. Try to organise your notes under these headings.

Apply:

You should apply your knowledge and understanding of polymers to specific uses. For example, explain why PET is often used to package fizzy drinks.

Examiner's tip

You can use the abbreviations for polymers instead of spelling out their difficult chemical names.

Topic 1.4: Glass

Recall:

You will need to know about glass used for commercial packaging.

Explain:

You need to know the properties, advantages/disadvantages and uses of glass. Properties can include aesthetic, functional and mechanical associated with glass. Try to organise your notes under these headings.

Apply:

You should apply your knowledge and understanding of glass to specific uses. For example, explain why glass jars are used to contain a variety of food products.

Examiner's tip

Most glass packaging is designed to resist impact and not shatter. Don't always think that the drawback of glass is that it smashes easily.

Topic 1.5: Woods

Recall:

There are three types of wood that you need to know about:

1. Jelutong – hardwood
2. Balsa – hardwood
3. Pine – softwood

Explain:

You need to know the properties, advantages/disadvantages and uses of the woods listed above when modelling and prototyping. Properties can include aesthetic, functional and mechanical associated with each material. Try to organise your notes under these headings.

Apply:

You should apply your knowledge and understanding of woods to specific modelling and prototyping uses. For example, explain why jelutong is a suitable material for making a vacuum-forming mould.

Examiner's tip

This is graphic products so woods used in furniture will not be examined – that is for RMT students only.

Topic 1.6: Composites

Recall:

There are two types of composite material that you need to know about:

1. Carbon fibre
2. Medium-density fibreboard (MDF)

Explain:

You need to know the properties, advantages/disadvantages and uses of the composites listed above. Properties can include aesthetic, functional and mechanical associated with each material. Try to organise your notes under these headings.

Apply:

You should apply your knowledge and understanding of composites to specific uses. For example, explain two properties of MDF that make it suitable for block modelling.

Examiner's tip

A great 'strength to weight ratio' is a major reason why carbon fibre is used in a wide range of products such as bike frames and racing cars.

Topic 1.7: Modern and smart materials

Recall:

There are five types of modern and smart materials that you need to know about:

1. Polymorph
2. Thermochromic liquid crystals/film
3. Liquid crystal displays (LCDs)
4. Electronic paper displays (EPDs) or e-paper
5. Transdermal prescription drug patches

Explain:

You need to know the properties, advantages/disadvantages, structural composition and uses of the modern and smart materials listed above. Properties can include aesthetic, functional and mechanical associated with each material. Try to organise your notes under these headings.

Apply:

You should apply your knowledge and understanding of modern and smart materials to specific uses. For example, discuss the use of polymorph when developing a design idea for a hand-held product.

Examiner's tip

Knowing the structural composition (the layers) of transdermal prescription drug patches will help you to explain why they are a great way of administering medication as opposed to injections.

Topic 1.8: Components

Recall:

There are four methods of binding that you need to know about:

1. Spiral/comb binding
2. Saddle-wire stitching
3. Perfect binding
4. Hard-bound or case-bound

There is a range of technical drawing equipment that you need to name and describe their use including:

Pencils, set squares, compasses, rulers, drafting aids (circle/ellipse templates, French curves/flexicurves) and drawing boards.

Explain:

You need to know the processes, advantages/disadvantages and uses of the binding methods listed above. Try to organise your notes under these headings.

Apply:

You should apply your knowledge and understanding of binding methods to specific uses. For example, explain why monthly glossy fashion magazines use perfect binding whereas weekly 'gossip' magazines only use saddle-wire stitching.

Examiner's tip

Questions relating to technical drawing equipment will feature strongly in the table in Q11a where you will either have to name or describe the use of that piece of equipment.

Know Zone
Chapter 2 Industrial and commercial processes

Topic 2.1: Scale of production

Recall:
There are three scales of production that you need to know about:

1. One-off
2. Batch
3. Mass

Explain:
You need to know the processes, characteristics, advantages/disadvantages and uses of these scales of production. Try to organise your notes under these headings.

Apply:
You may be asked in the exam to apply your knowledge and understanding of scales of production to a given use. You may also be asked about the social implications of using batch- and mass-production techniques which incorporate CAD/CAM and automated machinery. For example, discuss the impact on the workforce of the mass production of products.

Examiner's tip
Don't forget to revise the impact of batch- and mass-production processes on the modern workforce. These are not always positive.

Topic 2.2: Modelling and prototyping

Recall:
There are three processes that you need to know about:

1. Block modelling of MDF and Styrofoam™
2. Rapid prototyping using stereolithography (SLA)
3. Rapid prototyping using 3D printing (3DP)

Explain:
You need to know the processes, advantages disadvantages and uses of these forming techniques. Try to organise your notes under these headings.

Apply:
You may be asked in the exam to apply your knowledge and understanding of block modelling and rapid prototyping to a given use. You may also be asked to compare traditional modelling with rapid prototyping using CAD/CAM techniques. For example, explain **two** benefits of using rapid prototyping for developing a new product compared with traditional block modelling.

Examiner's tip
Refer to the properties of MDF and Styrofoam™ when answering questions on the benefits of using block modelling.

Topic 2.3: Forming techniques

Recall:

There are four thermoforming techniques that you need to know about:

1. Blow moulding
2. Injection moulding
3. Vacuum forming
4. Line bending

Explain:

You need to know the processes, characteristics, advantages/disadvantages and uses of these forming techniques. Try to organise your notes under these headings.

Apply:

You may be asked in the exam to apply your knowledge and understanding of forming techniques to a given use. For example, give **three** reasons why the casing of the mobile phone is injection moulded.

Examiner's tip

Always avoid the phrases 'quicker', 'faster' and 'cheaper' when explaining the benefits of using these thermoforming techniques to make products.

Topic 2.4: Joining techniques

Recall:

There are four types of adhesive that you need to know about:

1. Epoxy resin
2. Polystyrene cement
3. Tensol® cement
4. PVA

Explain:

You need to know the processes, advantages/disadvantages and uses of the adhesives listed above when joining like (e.g. acrylic to acrylic) and unlike (e.g. wood to metal) materials.

Apply:

You will be asked in the exam to apply your knowledge and understanding of adhesives for joining like and unlike materials. For example, which one of the following adhesives would be used to join . . . ., in a multiple-choice-style question.

Examiner's tip

Questions on joining materials are most likely to appear as multiple-choice questions in the exam.

Topic 2.5: Finishing techniques

Recall:

There are three techniques for enhancing the format of paper and board that you need to know:

1. Laminating
2. Varnishing
3. Hot-foil blocking

Explain:

You need to know the processes, advantages/disadvantages and uses of these finishing techniques. Try to organise your notes under these headings.

Apply:

You may be asked in the exam to apply your knowledge and understanding of finishing techniques to a given use. For example, give two reasons for laminating a menu in a restaurant.

Examiner's tip

Try not to confuse laminating with 'encapsulation' where you can see a sealed edge around the printed product.

Topic 2.6: Printing processes

Recall:

There are five printing processes that you need to know about:

1. Photocopying
2. Offset lithography
3. Flexography
4. Gravure
5. Screenprinting

Examiner's tip

You will have to revise a lot of specialist technical words in this topic in order to demonstrate a full understanding of each printing process.

Explain:

You need to know the processes, advantages/disadvantages and uses of these printing processes. Try to organise your notes under these headings.

You should use the internet to look for animated diagrams of these processes that will show you the process in use.

Apply:

You may be asked in the exam to apply your knowledge and understanding of printing processes to a given use. For example, explain two reasons why packaging is often printed using flexography.

Topic 2.7: Health and safety

Recall:

You need to describe safe working practices and how to identify hazards and precautions using risk assessments.

Examiner's tip

Health and Safety Executive (HSE) style risk assessments are key to answering these type of questions.

Explain:

The Health and Safety Executive's (HSE) 'five steps to risk assessment' should be used to:

1. Identify the hazard
2. Identify the people at risk
3. Evaluate the risks
4. Decide upon suitable control measures
5. Record risk assessment.

Ask your teacher or technician to photocopy some of the more relevant risk assessments that they have had to perform as part of your school's health and safety policy.

Apply:

You should apply your knowledge and understanding of health and safety to risk assessments of common design studio and workshop practices. For example, using a computer to design something and cutting a material using a vibrosaw.

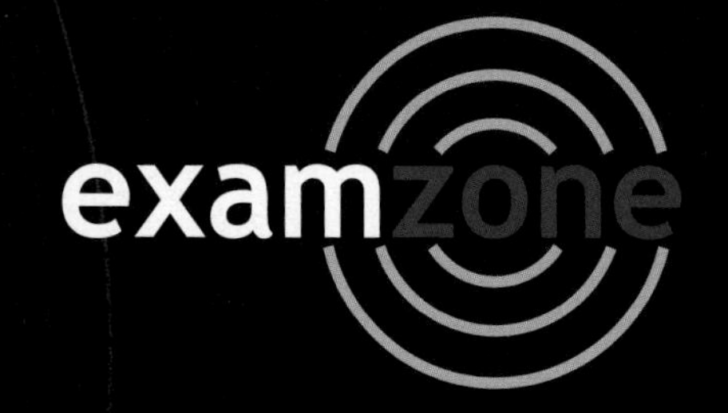

Know Zone
Chapter 3 Analysing products

Topic 3.1: Printing processes

Recall:

When analysing a product, you need to know about:

- **Form** – why is the product shaped/styled as it is?
- **Function** – what is the purpose of the product?
- **User requirements** – what qualities make the product attractive to potential users?
- **Performance requirements** – what are the technical considerations that must be achieved within the product?
- **Material and component requirements** – how should materials and components perform within the product?
- **Scale of production and cost** – how does the design allow for scale of production and what are the considerations in determining cost?
- **Sustainability** – how does the design allow for environmental considerations?

Explain:

Write the specification headings and their definitions on separate sticky notes. Mix them up and try to match the relevant specification heading with the correct definition.

Apply:

Look at a product that you are familiar with and use the specification headings to analyse it. Now do the same with a product that you don't know much about.

Examiner's tip

The product analysis question will always be Q13 so study the structure of this question carefully.

Topic 3.2: Materials and components

Recall:

You will need to know what materials and components are used in the manufacture of products, including:

- Properties
- Advantages and disadvantages
- Justification for choosing.

Explain:

Why are certain materials and components used in a product? You should be able to justify their choice by explaining properties and advantages/disadvantages.

Apply:

All of your knowledge and understanding of the materials and components in Topic 1 can be applied to the analysis of a wide range of graphic products.

Examiner's tip

The materials and components are those listed in Topic 1.

Topic 3.3: Manufacturing processes

Recall:

You will need to know what processes are involved in the manufacture of products, including:

- Stages in the process
- Advantages and disadvantages
- Justification for choosing.

Explain:

Why are certain manufacturing processes used to make a product? You should be able to justify their choice by explaining properties and advantages/disadvantages.

Apply:

All of your knowledge and understanding of the manufacturing processes in Topic 2 can be applied to the analysis of a wide range of graphic products.

Examiner's tip

The manufacturing processes are those listed in Topic 2.

Know Zone
Chapter 4 Designing products

Topic 4.1: Specification criteria

Recall:
When designing a product, you should take into account the following:

- **Form** – how should the product be shaped/styled?
- **Function** – what is the purpose of the product?
- **User requirements** – what qualities would make the product attractive to potential users?
- **Performance requirements** – what are the technical considerations that must be achieved within the product?
- **Material and component requirements** – how should materials and components perform within the product?
- **Scale of production and cost** – how will the design allow for scale of production and what are the considerations in determining cost?
- **Sustainability** – how will the design allow for environmental considerations?

Explain:
Write the specification headings and their definitions on separate sticky notes. Mix them up and try to match the relevant specification heading with the correct definition.

Apply:
In groups of three or four, each person writes a design specification like the ones in Q12 of the exam paper. Swap specifications with each other and try to design a product that meets the other person's criteria.

Examiner's tip
The product design question will always be Q12 so study the structure of this question carefully.

Topic 4.2: Designing skills

Recall:
You should be able to respond creatively to design briefs and design specification criteria, including:

- Clear communication using notes and sketches
- Annotation that relates back to the specification criteria.

Explain:
Do your sketches clearly communicate your design intentions? Could extra annotation explain something if your drawing skills are not as good as you would like? Have you checked that your annotation explains your design thinking and relates back to the specification criteria?

Apply:
In your groups, continue with the specification criteria activity above and sketch two ideas for the product which meet those criteria. Swap responses and use peer-assessment to mark each idea out of 8 marks.

Examiner's tips
Do not use light pencil to sketch as it might not be clear. Use a dark pencil or pen instead.

There is no need to use colour or shading as there are no marks available for the quality of your sketches.

Topic 4.3: Application of knowledge and understanding

Recall:

You need to know about a wide range of materials and manufacturing processes used for making a range of products.

Explain:

Check that your design ideas include annotation that refers to the choice of materials and processes. Why have you used those materials and processes?

Apply:

You should be applying relevant knowledge and understanding of the materials and processes you have studied to the design of a new product.

Examiner's tip

It is important to know about the materials in Topic 1 and processes in Topic 2 here.

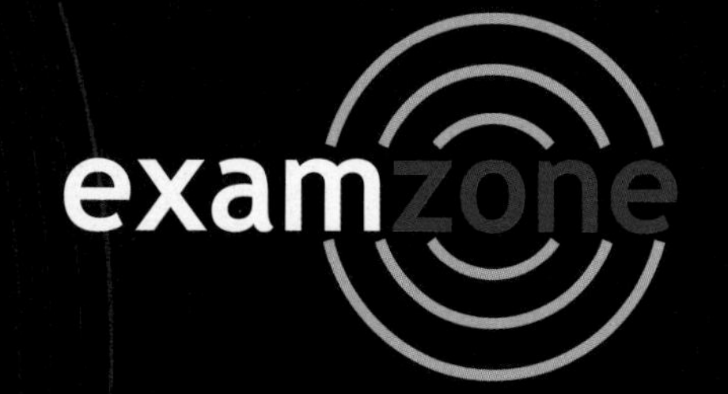

Know Zone
Chapter 5 Technology

Topic 5.1: Information and communication technology (ICT)

Recall:

There are four types of ICT that you need to know about:

1. Email communication
2. Electronic point of sale (EPOS) in retail and manufacturing
3. Internet in marketing and sales.

Explain:

You need to know the characteristics, processes, advantages and disadvantages and uses of these three types of ICT. Try to organise your notes under these headings.

Apply:

You may be asked in the exam to apply your knowledge and understanding of ICT to a given use. For example, describe the advantages, to the manufacturer, of using the internet to market products. Note that the advantages are for the manufacturer and not benefits to the consumer.

Examiner's tip

Don't forget that ICT has disadvantages as well as advantages. For example, email is a great way of communicating but it also has its problems.

Topic 5.2: Digital media and new technology

Recall:

There are two types of digital media that you need to know about:

1. High definition television (HDTV)
2. Commercial digital printing.

There are two types of new technology that you need to know about:

1. Bluetooth
2. Radio frequency identification (RFID) tags.

Explain:

You need to know the characteristics, processes, advantages and disadvantages and uses of these types of digital media and new technology. Try to organise your notes under these headings.

Apply:

You may be asked in the exam to apply your knowledge and understanding of digital media and new technology in comparison to an older technology. For example, explain the benefits of using RFID tags instead of barcodes when tracking shipments of products from a manufacturer to a retailer.

Examiner's tip

There are some tricky technical terms that you need to understand and revise for Bluetooth™ technology.

Topic 5.3: Computer-aided design/computer-aided manufacturing (CAD/CAM) technology

Recall:

There are three types of CAD that you need to know about:

1. 2D drafting
2. Desktop publishing (DTP)
3. 3D virtual modelling and testing.

There are two types of CAM that you need to know about:

1. Laser cutting/engraving
2. Vinyl cutting.

Explain:

You need to know the characteristics, processes, advantages and disadvantages and uses of these types of CAD/CAM. Try to organise your notes under these headings.

Apply:

You may be asked in the exam to apply your knowledge and understanding of CAD/CAM to a given use. For example, give two reasons why a manufacturer has used a laser to cut out the letters for an acrylic shop sign.

Examiner's tip

You should be able to link CAD/CAM with its impact on modern manufacturing techniques and the modern workforce in Topic 2.1 Scales of production.

Know Zone
Topic 6 Sustainability

Topic 6.1: Minimising waste production

Recall:
There are four ways of minimising waste production that you need to know about (4 Rs):

1. Reduce
2. Reuse
3. Recover
4. Recycle.

Explain:
You need to know the principles, advantages and disadvantages and uses of the 4Rs in minimising waste production. Carry out a life cycle assessment (LCA) for a product that you are familiar with, including: raw materials extraction, materials production, production of parts, assembly, use and disposal stages. How can the 4Rs be applied to each stage?

Apply:
You should apply your knowledge and understanding of the 4Rs to a range of graphic products and the materials and processes used to make them. You may also be asked to compare two similar products in terms of minimising waste production.

Examiner's tip
Cross-reference the 4Rs with the materials listed in Topic 1 and manufacturing processes listed in Topic 2.

Topic 6.2: Renewable sources of energy

Recall:
There are three sources of renewable energy that you need to know about:

1. Wind energy
2. Solar energy
3. Biomass/biofuels.

Explain:
You need to know the characteristics, advantages/disadvantages and uses of these three sources of renewable energy.

Apply:
You should apply your knowledge and understanding of these sources of renewable energy, especially their environmental impact in comparison to finite sources of energy such as oil, coal and gas.

Examiner's tip
Know the general benefits of using renewable sources of energy in comparison to non-renewable (finite) sources.

Topic 6.3: Climate change

Recall:
You need to know about:

- The responsibilities of developed countries in minimising the impact of global warming and climate change
- Reduction of greenhouse gases through the Kyoto Protocol.

Explain:
You need to know the principles of the Kyoto Protocol and the responsibilities of developed countries in reducing their greenhouse gas emissions. Why is it difficult for countries to meet their targets?

Apply:
You should apply your knowledge and understanding of climate change to other sustainability questions. For example, the benefits of using renewable sources of energy include the reduction of carbon dioxide and other greenhouse gases.

Examiner's tip
The Kyoto Protocol is a specialist area of the course and will not appear every year in the exam.

Know Zone
Chapter 7 Ethical design and manufacture

Topic 7.1: Moral, social and cultural issues

Recall:
There are three aspects of moral, social and cultural issues that you need to know about:

1. Built-in obsolescence as part of a 'throwaway' culture
2. Offshore manufacturing by multinationals
3. Cross-cultural design.

Explain:
You need to know the strategy, characteristics, advantages/disadvantages and uses of the above.

Apply:
You may be asked questions that will instruct you to discuss or evaluate these issues. Extended writing questions require you to mention both advantages and disadvantages in order to achieve full marks. For example, if discussing the use of built-in obsolescence when marketing new products, the issues are not always negative.

Examiner's tip
Don't forget to revise the disadvantages of offshore manufacturing to the local communities in developing countries as part of offshore manufacturing by multinationals.

Don't Panic Zone

Once you have completed the revision in your plan, you'll be coming closer and closer to the big day. Many learners find this the most stressful time and tend to go into panic mode, either working long hours without really giving their brain a chance to absorb information or giving up and staring blankly at the wall. Follow these tips to ensure that you don't panic at the last minute.

TOP TIPS

1. Test yourself by relating your knowledge to design and technology issues that arise in the news or by watching TV – can you explain what is happening in these issues and why?
2. Look over past exam papers, sample assessment materials (SAMs) and their mark schemes. Look carefully at what the mark schemes are expecting you to do in relation to the question.
3. Do as many practice questions as you can to improve your exam technique, help manage your time and build your confidence in dealing with different questions. Have a go at the activity questions in this book first.
4. Write down a handful of the most difficult bits of information for each topic on revision cards. At the last minute focus on learning these.
5. Relax the night before your exam – last-minute revision for several hours rarely has much additional benefit. Your brain needs to be rested and relaxed to perform at its best.
6. Remember the purpose of the exam – it's for you to show the examiner what you have learnt.

LAST MINUTE LEARNING TIPS FOR DESIGN AND TECHNOLOGY

- Remember that an intelligent guess is better than nothing. Graphic products – If you can't think of ways of minimising waste production then take a guess – you cannot lose marks.
- Know your topics – don't go into the exam unclear about basic definitions. Reduce, reuse, recover and recycle. Check out the Glossary.
- Many exam questions ask you to explain things. Make sure that you revise the skills that help you do this effectively, justifying all of the statements you make.

Exam Zone

Here is some guidance on what to expect in the exam itself: what the questions will be like and what the paper will look like.

UNIT	% OF OVERALL GCSE	MARKS	DESCRIPTION	KNOWLEDGE AND SKILLS
Unit 1 Creative Design and Make Activities	60	100	You must complete a design and make activity. These activities can be linked (combined design and make) or separate (design one product, manufacture another). **Task setting** You will choose from a list of five broad themes provided by Edexcel. **Task taking** This unit is internally assessed in controlled conditions. All work must be your own.	The assessment objectives covered in this unit are: Recall of knowledge and understanding **AO1**: 6% Application of knowledge and understanding **AO2**: 45% Product analysis **AO3**: 9% You will follow the basic creative design process. This includes research, product development, communication skills, application of knowledge and understanding of Graphic Products (materials, processes etc.), planning and making a high-quality product, testing and evaluating.
Unit 2 Knowledge and Understanding of Graphic Products	40	80	• The assessment of this unit is through an examination paper set and marked by Edexcel, lasting one hour and 30 minutes. • The examination paper will be structured in the same way each year so that it is accessible to all learners. • The examination paper will be a question and answer booklet and all questions are compulsory.The examination paper will consist of multiple-choice, short-answer and extended-writing type questions.	The assessment objectives covered in this unit are: Recall of knowledge and understanding **AO1**: 24% Application of knowledge and understanding **AO2**: 8% Product analysis **AO3**: 8% You will develop a knowledge and understanding of a wide range of materials and processes used in Graphic Products and important issues in Design and Technology. The knowledge and understanding that you develop in this unit can be easily applied to your Unit 1 Creative Design and Make Activities.

The examination paper is written to a template structure so that it is easy to prepare for. Each Graphic Products exam paper will be structured as follows:

QUESTIONS 1–10	QUESTION 11a	QUESTIONS 11b ONWARDS	QUESTION 12	QUESTION 13	QUESTION 14
10 multiple-choice questions	Completing a table by giving the missing names and uses of 4 different tools and pieces of equipment	Knowledge and understanding of graphic products Structured short-answer questions based on a theme	Designing products Use annotated sketches to design 2 different ideas for a given design specification	Analysing products Structured short-answer questions and one extended writing question based on a specific product	Knowledge and understanding of graphic products Structured short-answer questions and one extended- writing question
10 marks	**4 marks**	**15 marks**	**16 marks**	**16 marks**	**19 marks**
				TOTAL for paper	**80 marks**

ASSESSMENT OBJECTIVES

The questions that you will be asked are designed to examine the following aspects of Graphic Products. These are known as Assessment Objectives (AO). There are three AOs.

AO1	Recall, select and communicate your knowledge and understanding of Graphic Products plus its wider effects.
AO2	Apply knowledge, understanding and skills in a variety of contexts and in designing and making products.
AO3	Analyse and evaluate products, including their design and production.

THE TYPES OF QUESTION THAT YOU CAN EXPECT IN YOUR EXAM

The exam paper is 'ramped', which means that each question will get slightly more difficult as the paper progresses: Question 1 will be easier to answer than question 10, for instance. The advantage of ramping the whole paper is that the questions at the beginning of the paper are quite accessible, easing you into the exam and allowing you to gain confidence. As you work through the paper, the questions will get progressively more challenging. However, you are strongly advised to attempt all questions as there will be opportunities to gain marks throughout the paper.

This exam paper will contain several different types of questions:

- **multiple-choice questions** where you select the correct response from a choice of four
- structured **short-answer questions** which ask you to give/describe/explain your responses
- a **design question** where you will have to respond to a given brief
- **extended-writing questions** where you will have to evaluate/discuss/compare with longer responses.

UNDERSTANDING THE LANGUAGE OF THE EXAM PAPER

It is vital that you know what 'command' words ask you to do. Common errors are:

- *giving* two statements instead of *explaining* one
- *evaluating* without looking at both advantages and disadvantages.

Command word	Mark	Description
Give, State, Name	(1 mark)	These questions will usually appear at the beginning of the paper or question part and are designed to ease you into the question with a single statement or short phrase for one mark.
Describe, Outline	(2+ marks)	These type of questions are straightforward. They ask you to describe something in detail. Some questions may also ask you to use notes and sketches, so you can gain marks with the use of a clearly labelled sketch.
Explain, Justify	(2+ marks)	These questions will ask you to respond in a little more detail – single statements will not get you full marks. Instead, you will have to make a valid point and then go on to justify it to gain full marks.
Evaluate, Discuss, Compare	(4+ marks)	These questions are designed to 'stretch and challenge' you. They will be awarded the most marks because they require you to make a well balanced argument, usually involving both advantages and disadvantages.

Print your surname here, and your other names in the next box. This is an additional safeguard to ensure that Edexcel awards the right marks to the right candidate.

Ensure that you understand exactly how long the exam will last, and plan your time accordingly.

It is really important that you use black ink or ballpoint so that the Edexcel examiner can clearly see your responses.

Avoid using light pencil when sketching as it is not always clear to see.

Ensure that you read the instructions carefully.

You must answer all questions on this paper – there are no options.

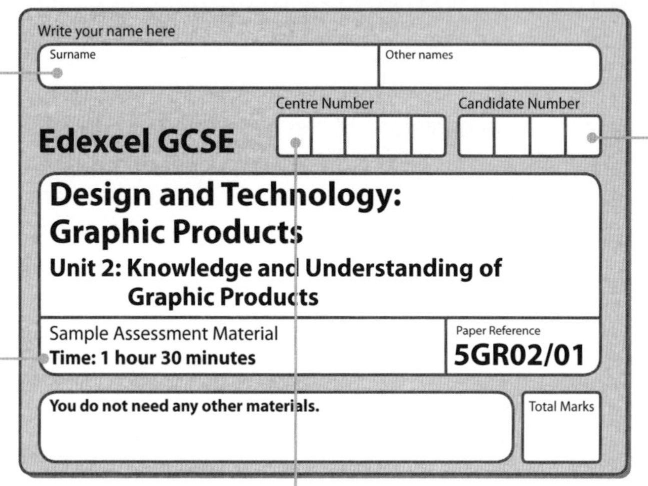
Write your name here

Surname | Other names

Centre Number | Candidate Number

Edexcel GCSE

Design and Technology: Graphic Products

Unit 2: Knowledge and Understanding of Graphic Products

Sample Assessment Material

Time: 1 hour 30 minutes

Paper Reference **5GR02/01**

You do not need any other materials.

Total Marks

Instructions

- Use **black** ink or ball-point pen.
- If pencil is used for diagrams/sketches it must be dark (HB or B). Coloured pens, pencils and highlighter pens must **not** be used.
- **Fill in the boxes** at the top of this page with your name, centre number and candidate number.
- Answer **all** questions.
- Answer the questions in the spaces provided *– there may be more space than you need.*

Information

- The total mark for this paper is 80.
- The marks for **each** question are shown in brackets *– use this as a guide as to how much time to spend on each question.*
- Questions labelled with an **asterisk** (*) are ones where the quality of your written communication will be assessed *– you should take particular care on these questions with your spelling, punctuation and grammar, as well as the clarity of expression.*

Advice

- Read each question carefully before you start to answer it.
- Keep an eye on the time.
- Try to answer every question.
- Check your answers if you have time at the end.

Turn over ▶

N37306A

1/6/4/4

edexcel
advancing learning, changing lives

Here you fill in your personal exam number. Take care when writing it down because the number is important to Edexcel in identifying who you are.

Here you fill in your school/college's centre number. You will be given this by your teacher on the day of your exam.

Note that the quality of your written communication will also be marked. Take particular care to use specialist technical terminology when answering questions with an asterisk.

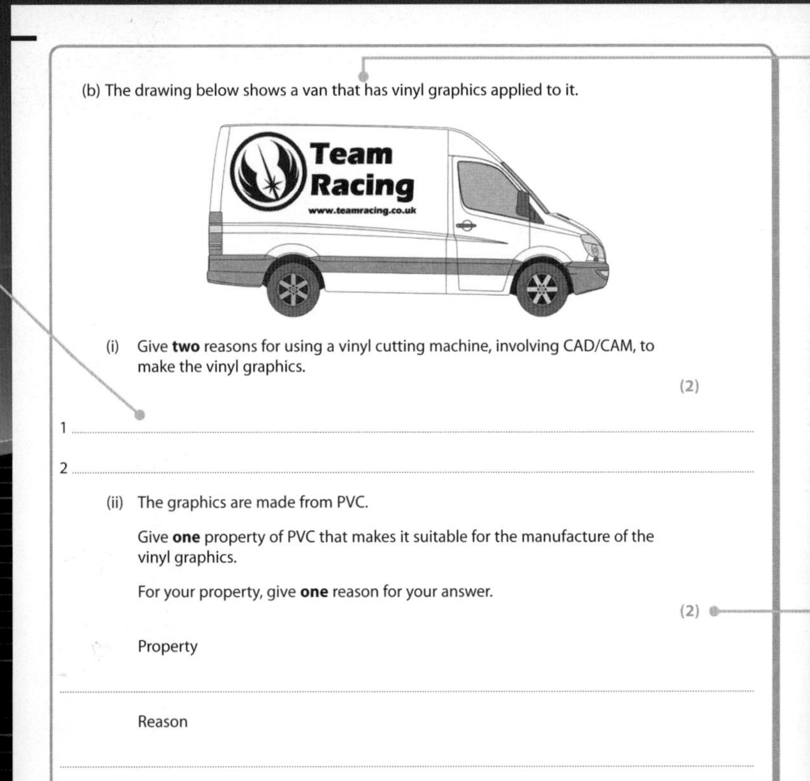

(b) The drawing below shows a van that has vinyl graphics applied to it.

(i) Give **two** reasons for using a vinyl cutting machine, involving CAD/CAM, to make the vinyl graphics.

(2)

1 ....

2 ....

(ii) The graphics are made from PVC.

Give **one** property of PVC that makes it suitable for the manufacture of the vinyl graphics.

For your property, give **one** reason for your answer.

(2)

Property

....

Reason

....

....

6

N 3 7 3 0 6 A 0 6 2 0

This first part of the question is called the 'stem'. It sets the scene for the questions that follow.

Often, stems will include pictures or diagrams to give you a feel for the context of the question

You must write your answers in the spaces provided.

Look at the amount of lines given as this indicates the length of your answer.

Marks are shown in brackets for each question.

Make sure you look at these before you start your answer as they indicate how many marking points you need to address.

Please do not write your answers in these blank spaces.

If you need to re-write your answer, use an additional sheet which can be attached to your exam paper.

Zone Out

Well done, you have finished your exam. So, what now? This section provides answers to the most common questions learners have about what happens after they complete their exams.

About your grades

Whether you've done better than, worse than or just as you expected, your grades are the final measure of your performance on your course and in the exam.

When will my results be published?

Results for your summer examination are issued in August. There are no January exams in this subject.

Can I get my results online?

Visit www.resultsplusdirect.co.uk, where you will find detailed learner results information including the 'Edexcel Gradeometer' which demonstrates how close you were to the nearest grade boundary. You can only get your results online if your school or college has given you permission to do so.

I haven't done as well as I expected. What can I do now?

First of all, talk to your tutor. After all the teaching that you have had, tests and internal examinations, he/she is the person who best knows what grade you are capable of achieving. Take your results slip to your tutor, and go through the information on it in detail. If you both think that there is something wrong with the result, your school or college can apply to see your completed examination paper and then, if necessary, ask for a re-mark immediately. The original mark can be confirmed or lowered, as well as raised, as a result of a re-mark.

If I am not happy with my grade, can I resit a unit?

Yes, you are able to resit each unit once before claiming certification for this qualification. The best available result for each unit will count towards your final grade.

What can I do with a GCSE in Graphic Products?

Graphic Products is well known as a subject that links to all other subjects of the curriculum, so a GCSE in Graphic Products is a stepping stone to a whole range of opportunities. A good grade will help you to move on to AS, Applied A Level, a Diploma or BTEC course. You may want to continue your study of Graphic Products by taking GCE Graphic Products or take a course such as a BTEC Diploma in Art and Design with a graphic design focus, which has a more work-related approach.

The skills that you develop can lead you to employment opportunities in advertising and marketing; a range of design disciplines including: graphic design, product design, interior design, vehicle design and architecture; and teaching.

Remember, every man-made object has been designed by someone for some need at some point in time. Therefore, there will never be a world without designers!

Glossary

A

Advertising: Any type of paid-for media that is designed to inform and influence existing and potential customers.

Aesthetics: The sensory responses that people make to stimuli such as colour, shape, smell and touch. Visual aesthetics refers specifically to those that are concerned with appearance i.e. the look of a product.

Automation: The use of machinery, rather than people, in manufacturing and data processing.

B

Barcode: Unique pattern containing a code to identify an item. Bar codes are the most common form of automatic identification used in EPOS to track retail goods.

Batch production: Manufacturing process in which components or products are produced in groups (batches) and not in a continuous stream.

Bauxite: The naturally occurring ore in the earth's crust that aluminium is made from.

Biodiversity: The variety of animal and plant species in an area.

Biofuels: Fuel sources derived from agricultural crops.

Biomass: organic matter such as timber and crops grown specifically to be burnt in order to generate heat and power or made into biofuels used in transport.

Blister packaging: Packaging in which the product is sealed in plastic, usually with a card backing.

Blow moulding: The process of using air, under pressure, to give form to hollow products using polymers e.g. PET fizzy drinks bottle.

Bluetooth: A radio-frequency (RF) based technology for wireless digital data communication over short distances.

Brainstorm: Hold a session to analyse key terms and suggest solutions to problems.

Brand: Unique design, sign, symbol, words used to create an image that identifies a product and differentiates it from its competitors.

Built-in obsolescence: Business practice of deliberately outdating a product before the end of its useful life and introducing a newer model or version.

C

Carbon neutral: Balancing carbon dioxide released into the atmosphere from burning fossil fuels by supporting renewable energy that creates a similar amount of useful energy, compensating for carbon dioxide emissions.

Cleaner design: Aimed at reducing the overall environmental impact of a product from 'cradle to the grave'.

Cleaner technology: The use of equipment and manufacturing processes that produce less waste or pollution than traditional methods. Therefore, reducing the use of raw materials, water, energy and lowering costs for waste treatment and disposal.

Climate change: Changes in global weather conditions, especially the increase in temperature over a period of time.

Components: The parts of a product that go to make up the whole.

Computer-aided design (CAD): The use of computer programs to design products and components with greater efficiency and accuracy.

Computer-aided manufacture (CAM): The use of computer controlled machinery and equipment to manufacture products and components with greater efficiency and accuracy.

Consumer society: A social culture in which consumers are encouraged by advertising to buy consumer goods.

D

Definition: How clear and distinct an image is.

Deforestation: The chopping down and removal of trees to clear an area of forest.

Design: Realisation of a concept or idea into a drawing, model, mould, pattern, plan or specification (on which the actual or commercial production of an item is based) and which helps achieve the product's specification.

Design specification: Sets out the criteria that the product aims to achieve.

Desktop publishing (DTP): The use of a computer program to design the arrangement of text, images and graphics on a printed page.

Developed countries: Countries at a late stage of development. They are generally rich, with a high proportion of people working in industry and services. Also known as More Economically Developed Countries (MEDCs).

Developing countries: Countries at an early stage of development. They are generally poor, with a high proportion of people working on the land. Also known as Less Economically Developed Countries (LEDCs).

Down time: Period of time when a manufacturing system is unavailable or offline.

E

Economies of scale: The cost of producing a single product or component falls when production increases.

Ergonomic: Designing products according to human needs, such as being comfortable and avoiding strain or injury.

Exploded drawing: Shows a product pulled apart, laid out in an ordered and linear form.

Finite resources: Natural resources of which there is a limited supply, like coal or gas.

Form: The shape and style of a product: its aesthetic properties.

Function: The means by which a product fulfils its purpose.

G

Gantt chart: A simple chart that maps each task against the time available, together with an order of priority.

Global market place: The marketing of products across the world.

Global warming: A trend whereby global temperatures rise over time, linked in modern times with human production of greenhouse gases.

Greenhouse gases: Those gases in the atmosphere that absorb outgoing radiation, hence increasing the temperature of the atmosphere e.g. CO2.

H

Hot-foil blocking: Finishing process used to print a metallic layer to paper and board.

I

Industrialisation: The process whereby industrial activity (particularly manufacturing) assumes a greater importance in the economy of a country or region.

Inert: Does not react chemically with materials and liquids.

Injection moulding: A highly automated manufacturing process in which a polymer is injected into a mould under high pressure. This process is particularly suited to producing complex shapes.

K

Kinetic energy: The energy an object possesses as it moves.

L

Lamination: Sandwiching layers of materials together.

Landfill: Disposal of rubbish by burying it and covering it over with soil.

M

Manufacturing specification: Clear details of product manufacture such as accurate drawings, clear construction details, dimensions etc.

Market: A place in the commercial world where buyers and sellers trade goods and services for money.

Market research: Identifies the buying behaviour, taste and lifestyle of potential customers.

Marketing: Anticipating and satisfying consumer needs while ensuring a company remains profitable.

Mass production: The manufacture of a product on a large scale using an assembly line, or another efficient means of production.

Modelling: Visualising design ideas using hand and/or computer techniques in 2-dimensions or 3-dimensions.

Modern materials: Materials that have been developed through the use of new or improved technologies, rather than naturally occurring changes.

Monoculture: Large scale production of a single crop.

Multinational: A company which does business in many countries.

N

Non-renewable sources of energy: Energy sources like coal or oil that will eventually run out.

O

Offset lithography: A widespread commercial printing process where ink is 'offset' from a printing plate to a rubber roller and then to paper.

Offshore manufacture: Assembly or full manufacturing in a country where labour and/or raw materials are cheaper, for export and/or eventual import into the manufacturer's home country.

One-off production: The manufacture of a single high quality product.

Outsourcing: A process in which a company subcontracts part of its business to another country.

P

Photovoltaic cell: A semiconductor that generates a small electric current when exposed to bright light.

Pixel: A small area of illumination on a display screen, from which pictures are formed.

Plasticity: The ability of a material to be moulded.

Point-of-sale: The place at which products are retailed.

Pollution: The presence of chemicals, noise, dirt or other substances which have harmful or poisonous effects on an environment.

Profit: The amount of money remaining from the selling price of a product after all costs of manufacture have been paid for.

Prototype: A detailed 3-dimensional model made from inexpensive modelling materials to test a product during the development process. A prototype can also be the first functioning product produced that many will follow form.

Q

Quality: Conformity to specifications and ensuring fitness-for-purpose. Making products right first time, every time, with zero faults to ensure customer satisfaction.

Quality assurance (QA): A system applied to every stage of design and manufacture ensuring conformity to specifications to make identical products with zero faults.

Quality control (QC): Checking and testing at critical control points in the manufacture of a product. Part of the overall quality assurance process.

R

Rapid prototyping (RPT): A CAD/CAM process that quickly creates 3-dimensional prototype products for testing during the development process.

Radio frequency Identification (RFID): Using electronic tags attached to products for storing data in order to identify and track them.

Renewable sources of energy: Sources of energy that can be re-grown or used again. They include wind and solar energy, biomass and sustainable/ managed forests.

Resolution: The degree of detail visible in an image.

Risk assessment: Identifying the hazards and risks to the health and safety of people in the workplace and putting suitable control measures in place to safeguard the risk of injury.

S

Scale of production: The size of production. See 'one-off', 'batch' and 'mass production'.

Smart materials: Can change in response to differences in temperature or light.

Solar energy: The light and heat generated by the Sun.

Solar panels: Used to gather solar energy from the Sun to generate electricity to produce heat.

Sustainability: The ability to keep our quality of life at the same rate or level. From this stems the idea that the current generation of people should not damage the environment in ways that will threaten future generations' environment and quality of life.

Sustainable resources: Resources, such as wood, that can be renewed if it act to replace them as it use them.

Sweatshop: A factory or workshop where manual workers are employed in poor conditions for low wages.

T

Target market group (TMG): All the customers of all the companies supplying a specific product.

Technical drawing: See 'working drawing'.

Technology: Purposeful application of information in the design, production and use of products, and in the organisation of human activities.

Tensile strength: A material's ability to resist stretching, compressing or pulling forces.

'Throwaway' culture: A culture with an attitude to consumption that pays little attention to the need to conserve resources.

Thumbnail: A small rough sketch showing the main parts of a design in the form of simple diagrams.

Tolerance: The degree by which a component's dimensions may vary from the norm and still be able to fulfil its function.

Translucent: Partially transparent, like frosted glass.

U

Unit cost: The price of manufacturing a single product or component.

V

Vacuum forming: A thermoforming process in which a softened polymer sheet is 'sucked' over a mould to make shell structures/casings.

Virtual reality modelling: Combines computer modelling with simulations to enable the development of an artificial 3-dimensional product or environment.

W

Webbing: Unwanted folds that can occur in the sheet of polymer material when vacuum forming.

Working drawing: Technical drawing, drawn at full size or to a scale and contains factual information relating to a product's appearance and dimensions. It should include all the necessary information for you or anyone else to make the product.

Index